HOME
SQUADRON

TITLES IN THE SERIES

New Perspectives on Maritime History and Nautical Archaeology

James C. Bradford and Gene A. Smith, editors

Rivers, seas, oceans, and lakes have provided food and transportation for humans since the beginning of time. As avenues of communication they link the peoples of the world, continuing to the present to transport more commodities and trade goods than all other methods of conveyance combined. The New Perspectives on Maritime History and Nautical Archaeology series is devoted to exploring the significance of the earth's waterways while providing lively and important books that cover the spectrum of maritime history and nautical archaeology broadly defined. The series includes works that focus on the role of canals, rivers, lakes, and oceans in history; on the economic, military, and political use of those waters; on the exploration of waters and their secrets by seafarers, archeologists, oceanographers, and other scientists; and upon the people, communities, and industries that support maritime endeavors. Limited by neither geography nor time, volumes in the series contribute to the overall understanding of maritime history and can be read with profit by both general readers and specialists alike.

HOME
─── ★ ───
THE U.S. NAVY ON THE
NORTH ATLANTIC STATION

SQUADRON

─────────── JAMES C. RENTFROW

Naval Institute Press
Annapolis, Maryland

Naval Institute Press
291 Wood Road
Annapolis, MD 21402

Library of Congress Cataloging-in-Publication Data

Rentfrow, James C.

 The U.S. Navy on the North Atlantic Station / James C. Rentfrow.

   pages cm

 Includes bibliographical references and index.

 ISBN 978-1-61251-447-5 (hardcover : alk. paper)—ISBN 978-1-61251-448-2
(epub) 1. United States. Navy. North Atlantic Squadron—History. 2. Naval tactics—History—19th century. 3. Operational art (Military science)—History—19th
century. 4. United States. Navy—Maneuvers—History—19th century. 5. Warships—
United States—History—19th century. 6. United States. Navy.—Reorganization—
History—19th century. 7. United States—History, Naval—To 1900. I. Title.

 VA63.N8R46 2014

   359.3'1—dc23

                                    2013050002

∞ Print editions meet the requirements of ANSI/NISO z39.48-1992
(Permanence of Paper).

Printed in the United States of America.

22 21 20 19 18 17 16 15 14   9 8 7 6 5 4 3 2 1

First printing

For

Sherman K. Lowe
Seaman 2/C, U.S. Naval Reserve, 1945–46

Jack Edwards
Chief Petty Officer, USN (Ret.)
U.S. Navy SEABEES, 1942–77
Kwajalein, Saipan, Guam, Vietnam (three tours)

CONTENTS

ACKNOWLEDGMENTS

In any academic endeavor, scholars depend on a host of supporters. This project was no different. I was honored to be the recipient of the Naval History and Heritage Command's Samuel Eliot Morison Supplemental Scholarship award for 2007. This work, I hope, justifies their investment in me.

The Navy/Maritime Records archivists at National Archives I in Washington, D.C., were especially helpful. They patiently fixed many incorrectly filled out pull requests and guided me in the hunt for more information. The staff of the Manuscript Division at the Library of Congress was similarly supportive. The U.S. Naval Academy's Nimitz Library in Annapolis is home to an unparalleled treasure trove of both primary and secondary naval history sources, and the staff there consistently goes above and beyond the call of duty, both for midshipmen and other researchers. Special thanks to History Department reference librarian Barbara Manvel for her consistent support. Dr. Evelyn Cherpak of the Naval War College's Naval Historical Collection was a gracious host during a research trip to Newport, Rhode Island. Janis Jorgensen of the U.S. Naval Institute's Photo Archive found every single illustration I asked for. I am especially indebted to Jesse Lebovics of the Independence Seaport Museum in Philadelphia for allowing me access to do some hands-on history, crawling around the USS *Olympia*.

The History Department of the U.S. Naval Academy is a great place to work. Department Chair Rich Abels was ever patient with me as this project took longer and longer to complete. I am particularly indebted to colleagues Lori Bogle, Nancy Ellenberger, Fred Harrod, Bob Love, and Aaron O'Connell for insightful comments on various chapters. Kate Epstein of Rutgers University, Camden, and Tim Wolters of Iowa State University both do great work on the U.S. Navy in this same time period and offered advice and support. I could not have asked for a better mentor than Dr. Jon T. Sumida of the University of Maryland, who was generous with his time and carefully read countless manuscript drafts. Any errors in interpretation or fact that remain are, of course, my own.

Finally, my ever-expanding family has been indulgent of Dad's frequent disappearances into the den. Thank you, Kathy, Melissa, Kaitlyn, Jonathan, Jeffrey, and baby Michael.

★

Introduction

This work examines the transformation of the United States Navy as a fighting organization that took place on the North Atlantic Station between 1874 and 1897. At the beginning of this period, the warships assigned to this station were collectively administered by a rear admiral, but were operationally deployed as individual units, each of whose actions were directed by their captains. By 1897 the North Atlantic Squadron, or "Home Squadron," as it was known, was a group of warships constituting a protean battle fleet—that is, an organized body moving and fighting in close order, which meant that the actions of the captains were directed by a commanding admiral. Its officers and sailors trained and conducted tactical exercises together, cruised to overseas ports together, socialized on liberty, and fought together at Santiago de Cuba in July 1898 during the Spanish-American-Cuban War. The reason for this change in form was a change in function. The objective of American naval power in the event of war shifted from commerce raiding to being able to engage and defeat hostile battle fleets.[1] At the same time, moreover, the basic matériel of navies was undergoing radical changes. In 1874 most of the U.S. Navy's inventory consisted of wooden cruising vessels. The first steel warships were authorized in 1883 and entered service between 1885 and 1889. These unarmored cruisers were followed rapidly by armored cruisers, then battleships. By 1897 the entire North Atlantic Squadron comprised modern warships.

That the squadron underwent important changes in the period 1874–97 is unquestioned. The modern battleships that confronted the Spanish navy in 1898 are proof that significant changes in strategic purpose and matériel took place. Historians have studied both of these aspects extensively. However, the process the squadron went through to effect these changes has received little attention. The development of a multiship fighting capability was more than simply a matériel problem. It involved the development of doctrine, tactics, and a hierarchy of command suited to the control of a complex fighting organization. Structurally, official change did not come until the designation of a North Atlantic Fleet in 1902, followed quickly by the consolidation of the North Atlantic Fleet and the South Atlantic Squadron into the Atlantic Fleet in 1906.

1

The North Atlantic Squadron's identity as a warfighting unit had changed long before this, however, having become a combat unit with the cohesion necessary to carry out combat operations at the squadron and fleet level in the late nineteenth century. This critical process of change was accomplished by rigorous exercise at sea. An inquiry into the nature of this process gives insight into the combat effectiveness of the U.S. Navy prior to the War of 1898, as well as the Navy's role in U.S. imperial aspirations.

Historiography

In the years immediately following the Civil War, the nation was preoccupied with Reconstruction and with the incessant drive of settlers and railroads westward across the North American continent. Overseas expansion and business opportunities were of secondary importance during these years, and as a direct result, no impetus existed to expend resources on a navy. While many contemporaries and historians alike simply dismissed the post–Civil War U.S. Navy as worthless, later historians have conceded that the wooden cruising vessels of the 1870s were exactly the equipment needed to carry out the Navy's primary mission of "showing the flag" and supporting U.S. business interests overseas.[2]

By the 1890s, this outlook had changed. Historians have debated (and continue to debate) at great length about the nature of this change—a "chicken or egg" argument about whether naval officers desiring a larger navy promoted expansionism or whether expansionist-minded politicians and civic leaders promoted a larger navy.[3] This study is concerned with operational rather than policy matters; it is not within the scope of this work to continue the policy discussion. Suffice it to say that, by the 1890s, with the North American continent conquered and the frontier "closed," a combination of desire for new markets and the "white man's burden" to share democracy and the American way of life with less-fortunate peoples turned the attention of Americans outward. This new outlook required a rejuvenated U.S. Navy.

The core histories of this naval revolution follow two basic lines of argument.[4] The first is theoretical and strategic in nature and covers the development of a new strategic purpose for the U.S. Navy. The standard narrative begins with Stephen B. Luce successfully agitating for the establishment of the Naval War College (NWC) in 1884. Luce then enticed Alfred Thayer Mahan to join the faculty of the new school. Luce and Mahan became the uniformed face of the so-called navalism movement. Together with politicians such as Theodore Roosevelt and Henry Cabot Lodge, they "called for a navy to fulfill the nation's destiny, and by 1890 agreed that it should be a 'blue-water navy'—a battle-oriented fleet of fighting ships."[5] This narrative typically culminates in the 1890 publication of Mahan's opus, *The Influence of Sea Power Upon History, 1660–1783*. In that work, Mahan argued decisively for the abandonment of the traditional U.S. naval strategy of coastal defense and commerce raiding and advocated a battle fleet that could protect American commerce and sea lines of communication.

The second line of argument emphasizes the political and legislative battles that led to the purchase and construction of the modern warships that made up the "New Steel Navy." This narrative typically begins with the Naval Advisory Boards of 1881 and 1882 and the authorization of the first four steel warships in 1883. It then traces the construction of various classes of ships, beginning with unarmored cruisers, then armored cruisers, and culminating with the introduction of battleships in the mid-1890s. The congressional battles to secure approval and appropriations for the various building programs are detailed. While some monographs treat only one of the main narratives,[6] most of the core histories address them both. The two lines of argument are deployed in arcs that intersect at the War of 1898, where the Navy's new strategic purpose and newly constructed warships are tested in combat.[7] Ancillary histories[8] explore structural and ideological aspects of the changes undergone by the naval officer corps[9] and the enlisted force,[10] or the changes wrought by new technologies.[11]

The traditional narratives are fundamentally incomplete. None addresses the crucial questions of the process of developing a multiship fighting capability.[12] With few exceptions, very little is said in any of these histories about the day-to-day operations of the Navy while engaged in this generation of transition from cruising to the battle line.[13] If operations are addressed, they are largely dismissed as a ragtag collection of ships haphazardly cruising around to various ports for the purpose of protecting American businessmen and their property.[14] Mahan and his battleships then arrive on the scene, sui generis, just in time to fight the battles of Manila Bay and Santiago de Cuba. This approach to U.S. naval history creates a "black box"[15] in which the wooden navy of the cruising era is entered in one end and the New Steel Navy magically appears from the other end in time for the War of 1898. The generation-long struggle of the operational Navy to re-create itself is entirely missing. This ignores the crucial development of the doctrine, tactics, and hierarchy of command necessary for a navy to possess a true multiship fighting capability. The organization that fought the war was not an inevitable outcome. The process of organizational change that produced a trained combat squadron of armored ships, and that squadron's combat effectiveness throughout the transformation, is the subject of this study.

It can be argued that tactical operations during this period have been rightfully deemphasized by naval historians who have grappled with larger questions of strategy and policy. However, history is an attempt to understand the thought processes of those who have gone before, to re-create not only their experiences but also how they conceptualized these experiences.[16] Thus it remains to look past the policy debates and examine the operational records of the Navy of 1874–97. In doing so, the historian can try to recover what it was that these naval officers were attempting to develop. Even though the ships they commanded were unarmored and had been designed for a naval strategy of commerce raiding, the fact that the Navy Department desired the capability to engage an enemy in line-of-battle formation, and repeatedly ordered this capability practiced whenever an opportunity presented itself, suggests imperial and expansionist tendencies in the United States prior to the 1890s. That critical decade in U.S. history can only be understood in the context of the preceding half-century.[17] To the extent that the North Atlantic Squadron was becoming less an administrative collection of ships and more an integrated combat unit, it can be argued that expansion and possible conflict with European powers was being conceived as early as 1874. In this way, an inquiry into the construction of the North Atlantic Squadron's identity provides evidence in the debate on the nature of imperialism and the United States.

I argue that evidence exists for America's "outward thrust"[18] in the day-to-day operations of the North Atlantic Squadron. The twenty-three-year span studied saw growing numbers of squadron exercises and attempts to solve the problems of maneuvering, short- and long-range signaling, basing strategies, coaling, and leadership. These problems had not been completely solved by 1898, but at least "ten years before Mahan," to borrow a phrase from Robert Seager, naval commanders were actively working to perfect fleet tactical maneuvering. It is easy to say that the construction of cruisers, commerce raiders, monitors, and other craft associated with an offensive defense in the 1880s is proof that those who were responsible for naval policy had no desire, prior to the authorization of the first battleships in 1889, for an overseas combat capability. It remains to address the operational record, however, and ask why these protected cruisers and commerce raiders were performing fleet tactics under steam and signaling exercises. Studying the operational record can uncover the middle ground between cruising responsibilities to protect American commerce, lives, and property and the uneasy relationship those responsibilities shared with professional naval officers' attempts to develop a military unit with fighting capabilities that would be useful against a peer European competitor.

The North Atlantic Squadron

As a unit of analysis, it is difficult to focus on the operations of all Navy warships around the world. For this reason, I considered it appropriate to select one of the squadrons as most representative of the change in the Navy as a whole. The North Atlantic Squadron, therefore, is the subject of this study. Among the cruising stations, only the North Atlantic Squadron was expected to carry out two distinct missions: the traditional cruising mission of protection of business and commercial interests abroad as well as the protection of the vital cities of the U.S. East Coast.[19] Its missions, therefore, most closely resembled European squadrons such as the

Royal Navy's Channel Squadron, which was equipped with the most modern matériel. Throughout the period studied, the North Atlantic Squadron received the latest equipment first, was close enough to the capital for the commander in chief to consult often in person with the Navy Department, and worked closely with the Naval War College to implement the latest thinking in tactics. The North Atlantic Station was considered to be a "plum" flag officer assignment. Often a prospective commander in chief would have already completed a CinC tour in one of the lesser squadrons before being rewarded with command of the coveted North Atlantic Squadron. A study of the process of change from a cruising navy to one that expected to fight in a battle line would rightfully start here.

At the close of the Civil War, the Navy Department moved to reestablish the overseas cruising stations that had been abandoned at the war's outbreak. Established by order of the Navy Department on the first of November 1865, the North Atlantic Squadron was formed by joining together the Atlantic Coast and West India squadrons. At the time of its formation, Secretary of the Navy Gideon Welles noted, "These squadrons . . . have, by one or more of their vessels, during the year visited nearly every principal port of the world. The views of the department enjoining activity, and the exhibition of the flag of our navy wherever our commerce penetrated, have been faithfully observed, and the reappearance of our men-of-war has been welcomed, not only by our countrymen, but by the people of every nation which they have visited."[20] The postwar national naval strategy is clear in the secretary's remarks. The Navy was, in the words of one junior officer, "absorbed in police duty for the State Department."[21]

The North Atlantic Squadron was responsible for two regions of intense U.S. foreign policy interest: the Caribbean and the Canadian fisheries. The squadron spent less time and devoted fewer warships to patrolling the fishing grounds off the coast of Canada than it did to the Caribbean. However, the touchy diplomatic situation surrounding fishing rights in the waters off Great Britain's Canadian colonies consistently threatened the otherwise-improving U.S.-British relations during this era. The rights of U.S. fishermen had been recognized as early as the Treaty of Paris, which ended the Revolutionary War in 1783. Although U.S. fishermen were no longer part of the British Empire, their livelihood rested in the cod fisheries off the coast of Newfoundland, and U.S. negotiators made recognition of this fact by the British a requirement for peace.

This recognition came in the form of Article III of the Treaty of Paris, which granted two things to American fishermen: the *right* to fish in the waters off Newfoundland and in all other international waters and the *liberty* to come ashore at uninhabited points along the Canadian coast to dry and preserve their catch. This last provision was just as important as the right to fish in the days before refrigeration and modern methods of getting a catch to market. Thirty years later, the War of 1812 caused confusion over the agreed-upon fishing rights, as the British government maintained that the Treaty of Paris had been nullified by the outbreak of hostilities in 1812. This necessitated a new agreement, the Fisheries Convention of 1818, which stipulated more precisely the exact geographic boundaries within which U.S. fishermen could both fish and approach the shore to dry and preserve their catch. It also forbade U.S. vessels from approaching any other harbors or settled areas not specifically authorized for any purpose other than to seek emergency shelter, repair damage, or purchase wood or water. The purpose was to prevent black-market trading between U.S. fishing vessels and the Canadian mainland. The final treaty between the two nations that addressed fishing was the Treaty of Washington, signed in 1871 to settle the Civil War–era *Alabama* claims. The Treaty of Washington reaffirmed rights for both American and Canadian fishermen, with an additional $5.5 million payment from the United States to Great Britain to compensate for what was judged to be greater concessions by the British.[22]

While the controversies over U.S. fishing rights in the Northeast occasionally kept North Atlantic Squadron warships busy patrolling the fishing grounds, it will be seen in this study that the majority of the commander in chief's time tended to be focused to the south. In the Caribbean region, three hotspots kept the squadron continuously busy. The island of Cuba, less than ninety miles away from mainland United States, had been on the mind of Americans since the Revolution. Generations of antebellum slave owners had coveted the island's land as a site for the expansion of slavery and the southern social and economic system. Postwar expansion enthusiasts were eager to reap the rewards of investment in the island's growing economy. At least two nineteenth-century administrations had attempted to buy the island outright from Spain but had been rebuffed. If the United States was unable to own Cuba, the next best alternative was a Cuba under the control of a weakened Spanish Empire, with a tacit understanding under the Monroe Doctrine that the island was not to be transferred to any other colonial power. As the Cuba Libre movement grew over the second half of the nineteenth century, the disorder caused by Spain's inept colonial governance threatened U.S. security and economic aspirations.[23]

Another island, Haiti, was a constant source of unrest. After its independence from France in 1804, the nation's mostly unstable government underwent a series of coups, as the military, elites, and commercial classes fought for control. Stability in Haiti was important to the United States for two reasons. By the 1870s, the search for overseas markets had led U.S. businessmen to Haiti, where U.S. property and investments were often in need of protection. More important, the island of Hispaniola contained excellent, and highly coveted, possible locations for naval stations. The U.S. government was acutely aware that whoever controlled these locations controlled the access to the Isthmus of Panama, with all its attendant commercial and national security implications. Haiti's

Môle St. Nicholas, site of Columbus' landing in the New World, was one such location. Repeated attempts by a succession of U.S. administrations to purchase or lease Môle St. Nicholas were unsuccessful, but the continual unrest in the nation kept the warships of the North Atlantic Squadron busy in and around Port-au-Prince.[24]

Finally, but perhaps most important, was the isthmus itself. With the expansion westward of the United States, the acquisition of California, and especially the discovery of gold in 1849, transit across the isthmus became critical to U.S. interests. Eventually a New York railroad company built a railroad across the isthmus, turning a four-day passage into a three-hour train ride and making the railroad and everyone financially associated with it exceptionally wealthy. Unfortunately for business investors, this was an area of great unrest. The citizens of Panama had attempted to secede from Gran Colombia several times since independence from Spain. The constant armed uprisings threatened not only the peaceful transit across the isthmus but also the property of the U.S.-controlled railroad company. Meanwhile, the attempt by the French builder of the Suez Canal, Ferdinand de Lesseps, to build a canal across the isthmus in the 1880s raised questions of European influence in the Western Hemisphere. The formal diplomatic relationship between the United States and Colombia was governed by an 1846 treaty that guaranteed the right of passage across the isthmus in exchange for U.S. guarantees of Colombian sovereignty. During the period of this study, U.S. troops would land at Panama twice and North Atlantic Squadron warships would be tasked to call at the port of Aspinwall almost continuously, under the terms of the 1846 treaty.[25]

Overview

What follows, then, is a narrative of the operations of the U.S. Navy's North Atlantic Squadron during the years 1874 to 1897. The selection of the time frame is deliberate. This is a study of operations amid changes in structure and organizational identity during an interwar period. The Civil War Union navy was a massive undertaking purchased and hastily built for the express purpose of combating the Confederacy and was largely dismantled within months of Appomattox. Other than understanding the very specific naval strategy that Abraham Lincoln employed to win the Civil War, the experience has very little to say about the strategic capabilities of the Navy beyond the lengthy southern coastline of the North American continent, or its desire to participate in overseas expansion. By the same token, this is not a battle piece about Manila Bay or Santiago de Cuba. Both engagements, and the War of 1898, have no shortage of historians eager to write about them in great detail.

Chapter 1, "The North Atlantic Squadron, the *Virginius* Affair, and the Birth of Squadron Exercises, 1874–1881," first examines developments in naval tactics in the mid-nineteenth century, introducing Commo. Foxhall A. Parker as the recognized U.S. expert in this area. It discusses the post–Civil War sea and shore organization of the U.S. Navy then turns to the 1873–74 concentration of the warships of the North Atlantic, South Atlantic, and European stations at Key West, Florida, during a war scare with Spain. When threatened with war, the Navy Department reacted in a manner that was exactly opposite the naval strategy it had embraced for public consumption. Rather than reinforce vital ports and prepare to sweep Spain's commerce from the seas through a robust program of commerce raiding, the Navy Department concentrated its wooden cruising warships and monitors, attempting to prepare for a multiship action against Spain's fleet. This series of exercises was so noteworthy that Rear Adm. Stephen B. Luce, assigned to the Lighthouse Board at the time, kept a copy of the handwritten journal of the fleet's movements with his personal papers.[26] This naval visionary recognized, as this study argues, that the maneuvers of the inefficient and obsolescent wooden warships marked the beginning of a process of significant change for the Navy. The future change in squadron identity is personified by Cdr. William Cushing, the commanding officer of the *Wyoming*. The month prior to the exercises, he and his ship had been one of the first responders to the capture of *Virginius*. While in Cuba, he had acted alone, on his own initiative, to protect U.S. citizens and property overseas. A month later, during the Key West exercises, *Wyoming* steamed second in a column of ships, with Cushing taking orders from the flag officer in command. It is a vivid illustration of the coming change in the cognitive experience of command, both for the commanding officers of the warships and the commander in chief.[27] In chapter 2, "Toward a New Identity, 1882–1888," the growth in the practice of holding squadron maneuvers is analyzed, providing evidence for the beginning of a change in identity for the squadron from an administrative organization to a combat unit. The early efforts of Rear Admiral Cooper to drill his wooden steam vessels are detailed, as are Rear Admiral Jouett's intervention in Panama and Rear Admiral Luce's handling of unrest in Haiti and in the Canadian fishing waters. The latter part of the chapter centers on Luce's innovative vision for a theoretical and operational partnership between the Naval War College and the North Atlantic Squadron. He had hoped to use his position as commander in chief of the squadron to complement his work at the Naval War College and develop the U.S. Navy's ability to fight fleet actions. The time had not yet come, however, for the primary mission of the squadron to be recognized as training and preparation for combat. The State Department still mandated the presence of U.S. warships throughout the North Atlantic Squadron's area of operations, and to his great disappointment, Luce was repeatedly unable to concentrate enough of his ships in one place to conduct tactical exercises.

Chapter 3, "The North Atlantic Squadron and the Squadron of Evolution, 1889–1891," discusses the acquisition and operational employment of the Navy's first four warships of the "New Steel Navy," the so-called ABCD ships. Although they were cruisers, the decision was made to operate them as a squadron—the "Squadron of Evolution." It was led by Rear Adm. John G. Walker, who had spent the previous eight years as the powerful chief of the Bureau of Navigation and Detail. Walker was a leader who displayed at once the understanding of what it meant to lead a squadron as a military unit and the limitations that flag officers raised in the "old Navy" faced when trying to adapt their style of leadership to the skills required in a new world shaped by steam propulsion and telegraph communications. Walker's role leading the Squadron of Evolution is contrasted with Rear Adm. Bancroft Gherardi's experience in command of the North Atlantic Squadron during the same time. While Walker led his squadron as a coherent unit, Gherardi was largely forced by various crises to manage his warships in the old-fashioned mode, detailing them throughout the Caribbean to carry out Navy Department tasking. Epitomizing the "warrior-diplomat" of bygone years, Gherardi became personally involved in diplomatic negotiations with the Haitian government. The limitations of the incompletely developed hierarchy of command for fleet operations were highlighted when the two squadrons met in Haiti in 1891.

In chapter 4, "The Limits of Ad Hoc Crisis Response, 1892–1894," Rear Adm. John G. Walker and Rear Adm. Bancroft Gherardi changed roles. Walker's Squadron of Evolution was broken up in late 1891, and he eventually became the commander in chief of the North Atlantic Squadron. He immediately had to cope with the Navy Department ordering his warships to various Caribbean ports as crises arose. Meanwhile, Gherardi was given command of the Squadron for Special Service and then the Naval Review Fleet, and his leadership style had to become more like Walker's as he struggled to develop the cohesion and unit identity necessary for those forces to carry out their missions successfully. Gherardi's commands provided more opportunity for the development of a concept of multiship operations but were limited in their ability to develop the doctrine and tactics necessary for a multiship fighting capability. Formation steaming for appearance's sake did not equate to formation steaming in combat. The chapter culminates with the 1893 International Naval Review in New York. Here, in an example of the effect naval pageantry could have on national identity and public opinion, the United States proudly displayed to the world—not to mention its own population—its new warships and administrative and operational prowess in handling large fleet operations.[28] The urbanization trend that led more and more Americans to live in or near the nation's large cities made it easier for large segments of the population to view the steel warships of the "new Navy."

Chapter 5, "Luce's Vision Realized: The North Atlantic Squadron Solidifies a New Identity, 1895–1897," describes how the squadron took the monumental step of deploying as a peacetime, noncrisis squadron. Rather than splitting up his forces for the customary winter cruises throughout the Caribbean in 1895, Rear Adm. R. W. Meade led the North Atlantic Squadron in visiting strategic ports throughout the region as a unit. While transiting from port to port, the squadron exercised regularly at fleet tactics and signaling. When in port, the officers mingled and the crews indulged in boat races. In their homeport of New York, squadron personnel organized a bicycle club and baseball teams. The extensive time spent steaming in company as well as the socialization of the crews of the squadron provide evidence that just prior to the outbreak of the Spanish-American-Cuban War, the North Atlantic Squadron had taken on a new identity as a combat unit. The squadron's function was not only to send single units to respond to State Department contingencies but also to be prepared to operate as a squadron in combat. Rear Admiral Meade's success as a squadron commander in chief was short lived, however. Unable to adapt his confrontational leadership to the consensus-building managerial style required by the new command functions, he was forced to resign his position after publicly insulting the president and secretary of the navy.

The epilogue narrates briefly the actions of the squadron during the Battle of Santiago de Cuba. While recognizing that the action at Santiago was not a true fleet engagement, it nonetheless provides a venue to reflect on the North Atlantic Squadron with respect to the development of a strategic purpose for a battle fleet, the development of a concept of multiship operations, and the test in combat action of a multiship fighting capability, as well as challenges yet to be overcome in the area of strong personalities and unified fleet command. While the North Atlantic Squadron had developed and demonstrated a protean combat capability, the process of becoming a battle fleet was incomplete. It awaited structural developments in the early twentieth century, such as the establishment of a fleet hierarchy of command. A battle fleet also required a concept of multiship operations that exercised a fighting capability rather than formation discipline geared toward appearances. Nonetheless, extensive and significant progress had been made toward the realization of this goal prior to the War of 1898. That progress is the focus of this study.

1

The North Atlantic Squadron, the *Virginius* Affair, and the Birth of Squadron Exercises, 1874–1881

Early on the morning of 4 February 1874, the U.S. naval forces on the North Atlantic Station[1] got under way from their anchorage off of Key West, Florida. Their mission was to execute fleet maneuvers under steam power.[2] As the assembled ships formed columns and steered to the southwest behind the lead ship, they represented the combined available combat power of the U.S. European, South Atlantic, and North Atlantic squadrons, the largest concentration of U.S. naval forces since the Civil War, which had ended nine years earlier. The series of exercises that took place over the next month has received relatively little attention from historians.[3] Nonetheless, the assembled ships were doing more than playing war games. The Key West exercises of February–March 1874 signal the beginning of the transformation of a cruising force into a battle fleet. Over the next twenty-three years, the squadron would undergo a series of core organizational changes, not only in matériel but also in terms of unit identity and the command technique necessary to direct a group of ships in battle. In the years 1874–81, the increasing desire on the part of the U.S. government and the Navy Department to possess a battle fleet created a conflict between the requirements of cruising missions undertaken by single warships and the need to concentrate and exercise warships in groups. Constant tasking in support of the State Department, "showing the flag" and protecting U.S. commercial interests, both abroad and domestically, regularly interfered with the North Atlantic Squadron's ability to train for combat against a peer naval power. Far from being a reactionary backwater period, the 1870s and 1880s were a time of considerable development, laying the essential foundation for the more visible naval buildup of the 1890s. During this time, naval authorities struggled to find the correct balance between traditional missions and the many changes in technology, matériel, and mission facing the North Atlantic Squadron.

Naval Tactics, 1874

Although European navies, as well as many progressive-minded U.S. naval officers, knew that the ability to maneuver warships in formation was critical to future naval combat, the exact nature of sound practice during a fleet engagement remained unclear. As ships became more heavily armored, neither the accuracy nor the explosive power of the guns of the 1870s was capable of penetrating the heaviest armor. The military problem to be solved at sea became one of penetrating the non-armored skin of an opposing warship below the waterline. The gun, which had been the featured weapon in the line-ahead battles of the eighteenth and early nineteenth centuries, was augmented by the ram and torpedo.[4] A "torpedo" in the mid-nineteenth century was simply an explosive device. Extended in front of the attacking ship on a spar, the torpedo constituted a ram with extended reach. The torpedo's explosive charge seemed to provide a much more accurate and controllable way of doing major damage to an ironclad vessel.[5] Torpedoes could be attached to a spar or, like the new British Whitehead torpedo, could be self-propelled. All of these represented options for a squadron commander in attack and defense.[6] Experts at the time were unsure how a large-scale battle would be fought. The U.S. Civil War offered little guidance because no action between fleets had occurred. One major naval action that took place not long after the Civil War, however, gave naval tacticians a rough idea of what combat between opposing groups of steam-powered ironclads would look like.

At the Battle of Lissa in 1866, an Austrian squadron made up of seven ironclads and fourteen unprotected vessels met an Italian squadron of ten ironclads and twenty-two unarmored units.[7] The Italian squadron, under the command of Count Carlo Pellion di Persano, was supporting a landing attempt on the island of Lissa, which was held by the Austrians. While shelling the island, they were attacked by the Austrian squadron, led by Wilhelm von Tegetthoff. The capital ships on both sides were of wooden construction and had full sail rigs, iron armor belts, and broadside-mounted armament. The Austrians employed a larger number of breech-loading guns than the Italians, although it was not to be the guns that would play the most dramatic role.

The Italians broke off their shelling of the Austrian positions and deployed for battle in a column formation. Tegetthoff, although outnumbered by the Italians, deployed his ships in a wedge formation and pressed home his attack, driving for the center of the Italian line. His intention, which he had expressed to his captains prior to the action, was to utilize the ram bows of his ironclads in a close action. He soon got his chance. On the Italian side, Persano had taken this inopportune moment to attempt to shift his flag, which threw his line into disarray and confused his commanding officers. The Austrian wedge broke through the line and a general melee ensued. What happened next influenced naval tactical thinking for the next thirty years. The lead ship of the Italian second division was *Re d'Italia,* an armored frigate displacing 5,700 tons and armed with six 72-pounder smoothbore shell guns and thirty-two 6-inch breech-loading rifles mounted in broadside.[8] She had a 4.5-inch armor belt. She was attacked by the Austrian flagship *Ferdinand Max*. Also an armored cruiser, she was slightly smaller than *Re d'Italia,* with sixteen 48-pounder, four 8-pounder, and two 3-pounder smoothbore guns. Her battery was protected by a 4-inch iron belt. Disregarding the rapid gunfire from the Italian breech-loading batteries, Tegetthoff executed ramming attacks on both *Re d'Italia* and another, smaller armored corvette, *Palestro*. Both Italian ships sank in minutes, at a cost of over eight hundred sailors, including both captains. The Italians eventually withdrew. Tegetthoff returned home to a hero's welcome, while Persano faced a court-martial and dismissal from the Italian navy.

From a strategic standpoint, the battle meant little. Italy and its ally, Prussia, prevailed in the conflict. The battle's impact on naval tactics had a wider effect. Attention focused on the ram as an offensive weapon of importance. The position of primacy of the gun was usurped, albeit briefly, by a return to tactics that seemed more suited to the age of galleys than the nineteenth century.[9] In one respect, bringing an enemy to action was now easier than it had been in the age of sail. Sailing ships relied on combinations of current and wind for propulsion and were only capable of sailing within six points (36 degrees) of wind direction. Station keeping in formation, not to mention closing with and engaging an enemy, were difficult propositions even for squadrons that exercised together regularly.[10] With the advent of steam propulsion, an admiral could command his formation regardless of wind speed or direction. Theoretically, ship's speed and turning radius were known quantities, as they could be measured and controlled by engineers. But as captains tried to position their ships to ram an opponent or launch a torpedo—both weapons that seemed capable of doing more lethal damage faster than guns—the neat line of battle was usually disrupted. Lissa cast doubt on the single column as a method of concentrating firepower.[11]

The development of the torpedo led to the adoption of small craft carrying torpedoes, which created new hazards for a line of battleships. In theory, it was possible that a small craft, cheaply built and lightly manned, could sink a capital warship. This created a dangerous paradox for the battle line. Concentrated by necessity to maximize the impact of its guns, the battle line was vulnerable to torpedo attack. To evade such attacks required the ability to shift course as a unit on demand—that is, the entire line of ships had to be able to move in a designated direction by carrying out a designated turn—making a naval officer's ability to handle his ship in close order and at high speed imperative for fleet action. Steam propulsion allowed fleet commanders to deal with these new threats to the battle line without being constrained by the direction of the wind, but that freedom came at a cost. Formations moved faster, and both commanders and subordinates had more decisions to make about direction and the disposition of formations and less time to make them. Successful accomplishment of these tasks required constant rehearsal at sea. While Naval Academy cadets received instruction in the theory of fleet tactics, the

1870s U.S. Navy did not possess the matériel resources necessary to both carry out its cruising mission and rehearse fleet tactics under steam.[12] The fleet concentration off Key West in late 1873–early 1874 gave the Navy Department a rare opportunity to address this issue operationally.

The U.S. Navy's foremost expert on the evolution of naval tactics under steam power was Commo. Foxhall A. Parker.[13] He is also one of the most underrecognized figures behind the Navy's shift from cruising to line-of-battle tactics. Much of this may have to do with the fact that he died unexpectedly at a relatively young age.[14] Parker entered the Navy as a midshipman in 1837. His development as a junior officer followed a standard antebellum path, with duty on board the *Constitution* serving under his father—a prominent naval officer in his own right—who was the commanding officer (CO). During the Civil War, he held several commands, one of which was the Potomac Flotilla, where he was responsible for protecting the Potomac River approaches to the capital and Alexandria, Virginia (held by the Union for most of the war). It was here that he gained experience participating in joint operations, as he had to work closely with the Army commander in the area. It was also here that he used his time in command of a number of small gunboats to work out the tactical evolutions that eventually would be published as the book *Squadron Tactics Under Steam* in 1864.[15] The end of the Civil War did not slow the output of his prolific pen. *The Naval Howitzer Ashore*, a short work on the use of artillery by naval battalions, was published in 1866. *Fleet Tactics Under Steam,* the sequel to *Squadron Tactics Under Steam,* came out in 1870. The Bureau of Navigation immediately provided four copies of this work for each ship fitting out for deployment.[16]

A "fleet" was defined as an assembly of twelve or more warships, arranged in three "divisions" that consisted of at least one "squadron" of no fewer than four vessels. Three basic formations covered all the options available to a fleet commander: line, column, and echelon (depicted in the figure).[17]

The influence of the 1866 Battle of Lissa can be seen in Parker's diagrams. The line (fig. 1) is identified as useful only for ironclads, rams, and torpedo vessels. The column is to be used only for vessels that mount broadside batteries. These were, as discussed above, the deployments used by Tegetthoff and Persano, respectively. As a result of Tegetthoff's success, the column fell briefly into disfavor as a formation for modern armored vessels. Improvements in gun technology and range would soon restore the column to primacy as a tactical formation, but Parker's diagrams represented the state of the art in 1869.[18]

In any of the formations, the vessels could take station at "open order," defined as two cable lengths, or "close order," a single cable length. A "cable" was 120 fathoms, or 720 feet; thus two cable lengths would be 1,440 feet, or just under a quarter of a mile. One cable length would be 720 feet, and "half distance" was 60 fathoms, or 360 feet. The ordered speed for a formation was at all times to be no greater than half a knot less than the maximum speed of the slowest vessel

Figure 1: The order of battle for *iron-clads, rams, and torpedo vessels:*

Figure 2: The order of battle for vessels whose fighting power is in their *broadside batteries:*

Figure 3: Orders offensive and defensive for *vessels of all* descriptions:

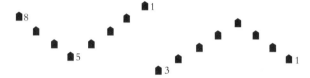

Vessels are said to be in direct echelon when steering the same course, each bears from its next astern, at an angle of 45% (4 points) from the course.

"This page from Commodore Foxhall Parker's Fleet Tactics Under Steam represents the state of the art in naval tactical thinking in the 1870's." Foxhall A. Parker, *Fleet Tactics Under Steam* (New York: D. Van Nostrand, 1870), pg. 10.

in the formation. This allowed the slowest vessel a reserve of speed available in order to gain and maintain station.[19]

The U.S. Navy in 1874: Structure and Organization

The Union navy during the Civil War boasted over seven hundred vessels of all types: steam-powered ironclads such as the famous *Monitor,* paddlewheel gunboats designed to operate in shallow rivers, and wooden cruising vessels designed for service on overseas stations. The large number of commissioned warships notwithstanding, there were no battles during the Civil War that required coordinated fleet maneuvers under steam power. This was mostly because of the small size of the Confederate navy, but it also had to do with the missions the Union navy was called upon to carry out during the war.[20] The few engagements that involved large numbers of ships were not fleet actions as such, but actually combined military operations that pitted Union ships against Confederate land fortifications and vessels that acted as extensions of those land defenses.[21]

After the close of hostilities, the U.S. Navy returned to its antebellum deployment patterns. The majority of the vessels constructed or purchased for wartime service were paid off and sold, and six stations were established: the North and South Atlantic, North and South Pacific, Asiatic, and European. By 1874 the entire U.S. Navy consisted of 163 warships. That number is misleading, as both Secretary of the Navy Robeson and Admiral of the Navy David Dixon Porter pointed out in their annual report to Congress that year, as it included tugboats and obsolete sail-only craft that were bound for sale and breaking up. The actual tally of effective warships was closer to seventy-three, which represented the number of fighting steam vessels suitable for overseas cruising. Even that figure, though, has a certain amount of ambiguity. Included in the count of seventy-three were several units that were on the docks or in various stages of repair or construction. A review of the warships actually assigned to the six cruising stations for the calendar year 1874 yields the following: The European Station had four vessels, the South Atlantic Station had one vessel, the South Pacific Station had three vessels, the North Pacific Station had six vessels, the Asiatic Station had eight vessels, and the North Atlantic, or "Home," Station counted ten cruising vessels and two monitors.[22] Thus the U.S. Navy had available the combat power of approximately thirty-four warships. To man and support these thirty-four warships, the Navy could draw upon an authorized strength of 10,000 officers and men and an annual budget of just over $23 million.[23]

The ships on these stations rarely worked together. Alfred Thayer Mahan noted in his memoirs that the "rule was that the vessels were scattered, one to this port, another to that . . . to the several officers their own ship was everything, the squadron little or nothing."[24] The organization of the Navy Department was also fragmented. Under the secretary of the navy, it was organized into eight bureaus by legislation in 1862: the Bureau of Navy-Yards and Docks, the Bureau of Construction and Repair, the Bureau of Equipment and Recruiting, the Bureau of Steam Engineering, the Bureau of Provisions and Clothing, the Bureau of Ordnance, and the Bureau of Navigation. This last was considered the first among equals, as it had cognizance over the Office of Detail, which controlled the movements of officers and ships. The eight bureaus oversaw a budget of just over $19 million. Cooperation among the bureaus was poor. Until the establishment of the Office of the Chief of Naval Operations after the turn of the twentieth century, there was no single operational commander between the bureau chiefs and the appointed civilian secretary of the navy. The admiral of the fleet—Adm. David Dixon Porter in 1874—was the principal uniformed advisor to the secretary of the navy, but his role was advisory. The exercise of executive authority was reserved for the secretary of the navy.

The Navy's infrastructure to service its warships on the East Coast was provided by six navy yards: Portsmouth, New Hampshire; Charlestown (Boston);

Philadelphia; Brooklyn, New York; Washington, D.C.; and Norfolk, Virginia. Significantly, there were no other naval facilities between Norfolk and the navy yard at Pensacola, Florida, on the Gulf of Mexico. This vexed station commanders, as they had to send their warships north, at least as far as Norfolk, if not farther, if they needed extended repairs or had yellow fever cases on board.[25] Navy dockyards, as major bases where ships were constructed and repaired, have received attention from historians.[26] Less written about are the smaller naval stations, which simply existed to resupply and recoal the Navy's vessels on the East Coast, but their role in enabling the vessels of the North Atlantic Squadron to concentrate would become increasingly important in the age of steam. Unlike sailing vessels, which once commissioned needed very little in the way of regular supply beyond fresh vegetables and meat, steam vessels required regular transfusions of coal and greater maintenance to the variety of engines and equipment carried on board. In the early days of the steam navy, the way this was dealt with was to have single ships pull into ports at their convenience and procure their own coal or supplies. This approach tied the number of vessels that could be serviced at any one time to the ability of the surrounding civilian work force and infrastructure to provide coal and supplies. For the fleet to spend more time concentrated as a fleet, it would be necessary to have a location that could coal and supply multiple ships; this was a subject of some concern to the Navy Department in the latter part of the decade.

Operationally, the problem with utilizing the existing navy yards as home ports lay within the Navy's bureau organization and structure. Under the regulations then in force, whenever a vessel was required to report to a navy yard, it came under the command of the commandant of that navy yard, meaning that the vessel was no longer under the operational control of the squadron commander in chief. This protocol often caused confusion regarding how to address correspondence and who should review it, as well as which organization (squadron or navy yard) was responsible for carrying out tasks associated with a particular warship.[27] This structural organization is evidence that the squadron commander in chief was not intended to concentrate his warships in one location as a routine practice. Instead, it was expected that ships that were in commission would spend the majority of their time deployed as single units, resupplying either off of the local economy wherever they were showing the flag or from one of the small Navy depots in the gulf region. The most often used of these for the North Atlantic Station was Key West, Florida, where the 1874 exercises would be held. But even at the depots, there were problems. The supplies of fresh vegetables and beef that could be had locally often could not supply more than one or two ships at a time.

The U.S. Navy in 1874: Innovation and Experimentation

The fact that U.S. warships were not concentrated or regularly rehearsing tactical maneuvers is not to say that there was no innovation happening within the

Navy Department in the 1870s. All major navies were experimenting with the torpedo during this decade, as previously discussed. Secretary of the Navy Adolph E. Borie recognized the importance and possible utility of this new weapon system. During his brief three-month tenure as President Grant's first navy secretary, he authorized the foundation of a torpedo school at Newport, Rhode Island.[28] While the Royal Navy was successful in fielding a self-propelled variant in the 1860s, U.S. naval officers were originally drawn to their use as stationary explosives, or what would be referred to as "mines" today. The thought was that strategically important U.S. harbors could be protected by a combination of monitors and a field of electronically detonated torpedoes.[29] Tactical, nonstationary versions were attached to the attacking ship either by a spar or towed alongside, physical forces holding the towed torpedo away from the offensive ship at a 45-degree angle until it could be positioned to make contact with an opponent.[30] Surprisingly, one of the biggest advocates of the torpedo ram was David Dixon Porter, the arch-conservative admiral of the Navy. History has left Porter with a reputation for resistance to change, especially for his infamous stinginess with coal and insistence on the use of sails for deployed warships, but in actuality Porter enthusiastically embraced certain innovations, and the torpedo boat was one of them.[31] The files of the chief of the Bureau of Ordnance are filled with letters from Porter—sometimes daily—excitedly talking about one of his pet torpedo projects.

One such enterprise involved *Alarm,* assigned to the New York Navy Yard to perform experimental work with torpedoes. She was 176 feet long and displaced 800 tons. Her main armament consisted of a 32-foot reinforced iron prow and a torpedo spar that extended from a watertight gland at the tip of the ram.[32] She would eventually be assigned to the Torpedo School at Newport, Rhode Island. *Alarm*'s weakness was the inability of her steam power plant to attain the high speeds necessary to mount a torpedo attack against a warship. At full steam, *Alarm*'s maximum speed was only ten knots.[33] Only slightly faster was the *Intrepid,* another torpedo ram of similar characteristics launched in Boston at about the same time. With a top speed of eleven knots, *Intrepid* was still considered too slow to carry out her designed mission of closing a large warship and maneuvering her spar torpedo into a position where it could be exploded under her waterline. Still, *Alarm* and *Intrepid* provided platforms from which the tactics and technology of the torpedo could be worked out for future generations of warfighters. These tactics had not yet progressed to the point where joint operations involving the torpedo ram and cruising vessels were contemplated.[34] Instead, most naval officers assumed that torpedo craft would be used in conjunction with moored torpedoes (mines) to guard harbors.[35] U.S. naval officers, particularly those on the European Station, watched developments in Europe carefully. Cruising warships were outfitted with Harvey (spar) torpedoes, which they practiced deploying regularly and reported the results to the Bureau of Ordnance.[36]

Regardless of the capacity in which torpedo boats would be utilized, it was apparent that flagships and/or larger fleet units would have to communicate with them. This, perhaps, is why the publication of new tactical codebooks in early 1869 was overseen by the Torpedo School, prior to the assignment of a chief signal officer (CSO) for the Navy.[37] This development took place in the summer of 1869.[38] The establishment of the office of the chief signal officer within the Bureau of Navigation was a tacit admission by the Navy of the fact that ships of the upcoming era would be required to work in company and communicate with each other far more often than occasional meetings in overseas ports. The future Navy was going to require tools and skills substantially more developed than exchanging numbers and requests to "send a boat on board" with the flagship every few weeks. Officers with specialized training in the art of signaling would be required, as would new and improved equipment. Several different systems of sending signals, both day and night, were experimented with during this time.[39] The amount of correspondence in the CSO's records dealing with methods of signaling between warships at night leads one to believe that the Navy Department assumed that future operations would require their ships to spend more time in formation.

Once the signal office was up and operating, Commo. S. P. Lee, the first chief signal officer, began to work on how best to get this important training to the fleet. The problem was attacked on two fronts. First, naval cadets at the United States Naval Academy in Annapolis were to be instructed in signals. By 1872 they were being trained in the naval system of signaling before graduation.[40] It worked, because later that year, commanding officers in the fleet were reporting to the Navy Department that the midshipmen received from the academy had been "thoroughly instructed at the U.S. Naval Academy."[41] At the same time, it was important to get training out to the fleet. This was accomplished by detailing an officer or two from each ship in commission to receive signals training at Washington, D.C. An agreement was made with the U.S. Army Signal Office to provide this training, as the Navy had decided to use the Army Code of Signals to transmit nontactical messages.[42] Once these core officers were trained, they were to return to their ships and provide training to the rest of the crew. The Quarterly Report of Signals Training was instituted by the office of the CSO to monitor each ship's progress.[43] Commanding officers were apparently annoyed at yet another piece of paperwork required by the Navy Department; the files of the CSO's office are stuffed with copies of complaints sent to the chief of the Bureau of Navigation about COs and their missing, incomplete, or sloppy quarterly reports.[44] One CO, however, was never late with his report and always made certain that his signal officers conducted their required training. Cdr. Stephen B. Luce, CO of the *Juniata* on the European Station, consistently filed the most comprehensive reports. It is easy, 140 years after the fact, to spot the many reports that were filled out at the last minute, with very little attention to the veracity of the report being made. Many commanding officers failed to grasp the importance of the coming changes in naval warfare, but Luce was not one of them.[45] It is evident that, as early as 1871, the future commander in chief of the North Atlantic Squadron and founder of the Naval War College was concerned about the details involved with training and fighting as a fleet rather than with a single ship.

Even with the amount of training being conducted, both at Washington, D.C., and afloat, the combination of naval tactical signals and the Army's signal book was unsatisfying to the Navy Department. In light of the changing face of naval warfare, it soon became evident that what was needed was a signal book that more closely complemented Commodore Parker's new book on steam tactics. In October 1872 a board appointed by the Bureau of Navigation was charged with "modify[ing] the Tactical Signal Book so as to conform to what is necessary for the execution of maneuvers . . . for Parker's Steam Tactics."[46] Captain Parker put in his own suggestions for the signal book as well.[47] By early 1873, the new signal book was ready for the fleet.[48] It was this new book that Commodore Parker—by then the new CSO—brought with him to exercise with the assembled vessels at Key West in early 1874.

The *Virginius* Affair

The U.S. Navy had mustered all the ships available in the Atlantic in Key West, Florida, in late 1873 in response to the threat of war with Spain. Long a declining imperial power, Spain's increasing inability to maintain order in its overseas possessions had caused uneasiness around the world.[49] In no location was this truer than Cuba, one of the focal points of the North Atlantic Squadron's operations in the Caribbean Sea.

The events that nearly led to war with Spain began in 1870, at the pier of the Washington Navy Yard, where a surplus former Confederate blockade runner named *Virgin* was moored. *Virgin* had been in and out of government hands since being captured by the Union navy during the war. A side-paddlewheel steamer, 225 feet long and with a powerful engine, she was offered for sale by the government and bought by an American businessman who was acting as a proxy for Cuban insurrectionists. They intended to use the ship to run troops and supplies to their positions in Cuba. Over the next three years *Virginius,* as she had been renamed, made several successful runs with weapons, supplies, and fresh troops. Along the way, the ship became well known to just about everyone in the Caribbean. It was no secret that she was supplying the Cuban rebels. U.S. naval forces in the Caribbean were familiar with the nature of work that *Virginius* was engaged in.[50] She was, however, flying the U.S. flag, which for a time stopped Spanish officials from seizing her.[51]

In October 1873 her luck ran out. After a chase of several hours, the *Virginius* was captured by a Spanish gunboat as she attempted to bring a load of war matériel and volunteers to the rebel forces in Cuba. The ship was brought into port at Santiago de Cuba, where the Spanish military authorities rushed through a perfunctory court-martial and sentenced her captain, Joseph Fry (who was an American citizen, Annapolis graduate, and former Confederate naval officer), as well as another fifty-three passengers and crew, to death. These executions were carried out, over the strenuous objections of the U.S. vice consul on the scene, between 4 and 13 November 1873. Fortunately for the rest of the imprisoned crew of the *Virginius,* a Royal Navy ship, HMS *Niobe* (4), Captain Sir Lambton Loraine commanding, arrived on the scene, having been dispatched by the admiral commanding the British West Indian naval forces. The British were at first reluctant to become involved in the filibustering imbroglio, but the U.S. consul at Jamaica managed to convince the admiral that there were British citizens on board the *Virginius.* Loraine intervened with the Spanish authorities to prevent any further executions. With British imperial disfavor now plainly displayed by a black-hulled warship at anchor in the harbor, Captain-General Burriel relented. The executions stopped.[52]

Meanwhile, the closest U.S. ship to Santiago de Cuba was the *Wyoming* (6), Capt. William Cushing commanding. She had arrived at Aspinwall, on the isthmus of Panama, in September 1873.[53] Operating independently, as was standard for U.S. ships on the North Atlantic Station, her mission was to protect U.S. interests in this always-important geographic nexus for travel and commerce. A recent revolution in Panama had made the U.S. merchants and railroad agents uneasy, and they welcomed the visit of a U.S. warship to their harbor.[54]

Cushing epitomized the independent-operating naval officer. A member of the Annapolis class of 1862, Cushing proved his stubbornness and determination before he even received a commission. When he was forced to leave the Naval Academy in his senior year due to academic difficulties, Cushing pushed for and received a wartime commission as a master.[55] Master Cushing excelled at his early wartime assignments and eventually won reinstatement to the rank of passed midshipman. Thanks to the exigencies of the Civil War, he was promoted quickly and soon was a lieutenant.

Cushing's exploits while on blockading duty off the coast of North Carolina represented a kind of leadership that the industrial revolution was rapidly making obsolete. Cushing acted independently, carrying out what he perceived to be his duty with minimal input from superiors and the freedom, as a relatively junior officer, to determine his own best course of action given whatever circumstances he was in. Often, this required him to act decisively first and seek approval from higher-ups after the fact. Cushing relished this role. "A ship at sea is a complete system in itself," he wrote his cousin. "The captain is king, and as absolute a monarch as ever lived. The officers are his house of lords, and some five hundred men are his subjects. . . . I had rather be an officer on board a man o' war than the President of les Etats-Unis."[56]

Cushing is best known for his daring raid on the Confederate ram *Albermarle* during the Civil War. *Albermarle,* constructed in a shipyard in North Carolina in 1863–64, was 158 feet long, carrying on her superstructure a casemate 60 feet long covered with 4 inches of iron plating. Her armament consisted of broadside guns and, more important, an armored bow that was to be used as a ram. She got under way on 17 April and made her way down the Roanoke River. She ran a gauntlet of Union fire at Warren's Neck, the shots from the Union fort bouncing off her casemate, and then happened upon two Union navy gunboats. Like most of the hurriedly purchased and armed vessels that the Union navy used for river patrol, *Miami* and *Southfield* were heavily armed but unarmored, and no match for the Confederate ironclad. *Albermarle* attacked the two boats aggressively, plunging her armored bow deep into *Southfield's* side and causing the gunboat to sink almost immediately. *Miami,* meanwhile, fired furiously into *Albermarle* at almost point-blank range, but the shots simply bounced off the Confederate ram's armored sides. Frustrated, and with her commanding officer disabled, *Miami* withdrew. Thus, in the space of a few minutes, *Albermarle* had established Confederate naval supremacy on the Roanoke River. After recovering a few survivors from *Southfield,* she returned to port. Another, larger engagement took place on 5 May. While *Albermarle* and another Confederate steamer, *Bombshell,* were escorting a troop transport, they were engaged by four Union warships: *Miami, Mattabesett, Sassacus,* and *Wyalusing. Bombshell* was forced to surrender, but *Albermarle* successfully traded fire with all four Union ships before retiring unharmed. The situation for the Union was now critical. The Union navy's ability to conduct naval warfare on the Roanoke River was completely negated by one armored ship, about which its gunboats could do nothing. The ram would have to be cut out and destroyed.

Unfortunately for the Confederate navy, Commander Fusser, the commanding officer of the *Southfield,* had been a close friend of Lieutenant Cushing. "I shall never rest until I have avenged his death," Cushing swore.[57] Without hesitation, he volunteered to take two small boats upriver to destroy the Confederate ironclad. The mission was carried out on the night of 27 October 1864. Leading two small boats and twenty men, he crept up the river, past a detachment of sentries standing watch on the sunken hulk of the *Southfield,* and drove a spar torpedo underneath the hull of the *Albermarle,* personally detonating it when it was placed correctly. He and his crew then dove overboard amid heavy small-arms fire and swam away. For his actions, Lieutenant Cushing was voted the thanks of Congress (which resulted in his being advanced in rank) and a large share of the prize money awarded by a court of admiralty when the hulk eventually fell into

Union hands. Northern newspapers made much of the brilliant naval exploit, at a time when civilian morale and zeal for prosecuting the war was at a low point in the Northern states.[58]

This then was the officer who received a telegram in November 1874 while at Aspinwall, Colombia, stating that American citizens were in danger in Cuba. Any naval officer of that era would be expected to act exactly as Cushing now did—taking immediate, decisive action. Although in this case Cushing's actions were not as immediate as the vice consul in Santiago de Cuba would have liked them to be. As much of a reputation as Cushing might have had for celerity, he does not seem to have been in much of a hurry to leave Aspinwall. This may have had something to do with the fact that everyone in the U.S. Navy knew exactly what *Virginius* was doing, and it was not exactly surprising that she had been caught in flagrant violation of the law. The assumption was that the U.S. flag flying from *Virginius'* mast would protect her passengers and crew. After receiving the initial telegram, Cushing telegraphed back to the vice consul at Santiago de Cuba, asking for "more information," then waited over the weekend (the telegraph office was closed) for his reply. Satisfied that American lives and property were, in fact, in danger, Cushing took on coal and departed from Aspinwall on 11 November.[59] After leaving Aspinwall, Cushing touched at Kingston, Jamaica, where he satisfied himself that the *Virginius'* papers were in order, then proceeded immediately to anchor at Santiago de Cuba.[60] Here Cushing met with Captain Loraine and the U.S. counsel. He then wrote a scathing letter to the Spanish military officer in charge, demanding both an immediate halt to the executions and a personal meeting, and threatened to fire on the town.[61]

Cushing met with General Burriel the next day. Hagiographic accounts depict Cushing standing with his "hand on the hilt of his revolver" making demands of the general while telling him to evacuate the women and children from Santiago before he shelled the city.[62] Other authors do not present such a heroic picture. Depending upon which account one reads, Cushing was either the single-handed savior of the *Virginius* crew, who took charge of events from an ineffectual British officer, or he arrived after the heavy lifting had been done by Captain Loraine and took the credit.[63] As is often the case in history, the truth probably lies somewhere in between. In any event, the combination of British and U.S. naval presence convinced the Spaniards that there was little to be gained by continuing to execute prisoners. The Spanish military officers had almost certainly heard of Cushing's Civil War exploits and perhaps took his threat to shell the city of Santiago de Cuba seriously.[64] The USS *Juniata,* Captain Braine commanding, soon arrived on the scene, and *Kansas* and *Canandaigua* followed as quickly as they could. As the senior officer present, Captain Braine took charge of things in Santiago and sent Cushing north with *Wyoming* to report on the situation in person. There is some evidence that the Navy Department was concerned that the hotheaded Cushing would start a war and had hurried Braine down to Cuba to relieve him as quickly as possible.[65]

During a meeting of President Grant's cabinet on 14 November, Secretary of the Navy Robeson was told to gather all available naval forces at Key West in anticipation of trouble with Spain.[66] Telegrams began to fly, as the cruising forces of the North Atlantic, South Atlantic, and European squadrons began to move toward Key West, Florida. It was decided that Rear Adm. A. Ludlow Case, the commander in chief of the European Station and senior to the North Atlantic Squadron's commander in chief, Rear Adm. Gustavus Hall Scott, would take command of the combined forces.[67] Case's chief of staff would be the Navy Department's current chief signal officer, Commo. Foxhall Parker. Parker's involvement was important because he had authored the Navy Department's newly published manual for steam tactics and brought copies of the new signal manual with him to Key West. In Case's words, "Some of the drills and exercises, directed by the department, which had commenced before his arrival, will now be carried on more efficiently, as he brings books for details to be followed by all vessels—a great desideration."[68] Parker's work up until now had been largely theoretical and based on the publications of foreign authors. This would be the Navy's first opportunity to have enough ships in one location to test the tactical formations that Parker had written about.

The crisis, however, ended before the fleet could be gathered. After hurried negotiations in both Madrid and Washington, D.C., the release of the prisoners was eventually agreed upon. *Juniata* took custody of 102 prisoners and departed Santiago de Cuba on 18 December 1873.[69] In this case, the mutual desire of both Spain and the United States to avoid war had prevailed. The limitations of the U.S. warships have already been discussed. The Spanish fleet of 1873 was superior in numbers and technology to the U.S. Navy. It boasted seven heavy armored vessels, ten screw frigates, three armored turret vessels, and five screw corvettes, as well as several advice vessels and gunboats.[70] European powers, propelled by defense requirements more pressing than those of the United States, invested early in such things as armor and improved armament. The Spanish ironclad *Arapiles*,[71] which was in dry dock at the Brooklyn Navy Yard at the time the *Virginius* affair took place, was an example of this. Spain was one of the few second-rate naval powers to invest in broadside ironclads like *Arapiles*.[72] Launched in 1864, *Arapiles* was 280 feet long and displaced 5,500 tons. Her main armament consisted of twenty-two 10-inch muzzle-loading rifles (MLR), five 8-inch MLR, and ten 7.9-inch breechloaders (BL) mounted in broadside. These were protected by a 4.25- to 4.75-inch armor belt of iron amidships. With a top speed of twelve knots, *Arapiles* was more than a match for any warship the North Atlantic Squadron could send against her. But she had suffered major damage off the coast of Venezuela while on deployment. As Spain had no dock facilities in its overseas possessions capable of repairing *Arapiles,* she was forced to put in at New York to have the work done and, ironically, was docked there during the 1873 diplomatic standoff with Spain. Although there were no problems during her detainment at the height of the *Virginius* crisis, her officers and crew wasted few opportunities to remind their hosts at the Brooklyn Navy Yard that she could outgun anything they owned.[73] The fact that a nearly obsolescent warship was widely viewed in the press as being superior to anything that the U.S. Navy could mount against it certainly gave U.S. policymakers reason for pause.

However, Spanish circumstances were also difficult. The Spanish Empire was mired in a costly civil war. Followers of Carlos VII controlled the city of Cartagena, where many of Spain's naval assets were located, and it was doubtful that the Spanish government would have access to them in the event of war.[74] On the whole, it was mutually beneficial to come to an agreement regarding the *Virginius.*

The account of the *Virginius* incident to this point highlights the performance standards that informed the professional prospective of a mid-nineteenth century U.S. naval officer. When the first telegrams from the panicked vice counsel at Santiago reached him at Aspinwall, Commander Cushing took immediate action. He had no opportunity to consult with Rear Admiral Scott, the commander in chief of the North Atlantic Station. He did not receive orders from any superior. The commanding officer of the *Wyoming* took it upon himself to get under way and steam for Jamaica, knowing that he was expected to deal with the situation. For Cushing, command involved making decisions that amounted to the execution of national policy at an operational, and even strategic, level. Cushing's performance in this episode epitomizes the warrior/diplomat naval officer of the first half of the nineteenth century. However, two months after the *Virginius* incident, the same USS *Wyoming* that had taken a city under her guns and risked war with the Spanish Empire was steaming in close order formation with eleven other warships. The skills that Commander Cushing would be called upon to display over the next two weeks had nothing to do with diplomacy or making decisions that would affect national policy. Instead, the *Wyoming* would be expected to respond to signals quickly and accurately, keep her station in close order with the ships in front and behind of her, and execute complicated maneuvers in concert with the other ships of the fleet while an admiral made strategic and operational decisions.

The 1874 Squadron Exercises

Rear Adm. A. Ludlow Case arrived in Key West on 3 January 1874.[75] Tapping Case to command the combined European, North Atlantic, and South Atlantic squadrons was an excellent choice. Case had entered the Navy as a midshipman in 1828.[76] After distinguished service in both the Mexican-American War and the Civil War, he spent the four years from 1869 to 1873 as the chief of the Bureau of Ordnance. Perhaps no other officer on active duty at the time was more familiar with the new weapons systems being deployed by European navies, and certainly no officer was more eager to experiment with new weapons and tactics, unless it was his chief of staff, Commo. Foxhall A. Parker.[77] The selection of a command team that featured an ordnance expert assisted by a tactical and signaling expert was one of the keys to success of the exercise.

Commodore Parker arrived in Key West on 22 January.[78] One of his first duties was to act as the commander of a naval brigade that carried out landing exercises on 23 January 1874.[79] Parker was the perfect man for this duty as well, having published works on the landing and employment of naval howitzers using small boats.[80] It is noteworthy that, in the midst of this unprecedented effort to exercise U.S. Navy warships tactically as a coherent combat unit capable of engaging an enemy squadron, Rear Admiral Case's first thoughts, before the exercises even got under way, were to practice landing troops. He understood that the primary mission of the U.S. Navy still involved protection of American interests on overseas soil.

Mostly, they were just waiting for all the ships that had been ordered to Key West to show up. It was a point of some embarrassment to the Navy Department that the alarm had gone out on 14 November, the crisis had ended in December, and it was not really until the end of January 1874 that anything approaching a "fleet" had been assembled.[81] By 4 February 1874, the twelve wooden cruisers and five ironclad monitors were in place and the exercises were ready to commence. The steaming characteristics of the monitors and the wooden screw vessels were so dissimilar as to make it almost impossible for the two ship types to operate together, so one of Case's first decisions was to exercise the two types of vessels separately.[82] The cruising vessels went first.

The North Atlantic Fleet's tactical drills evolved in a logical progression, designed by Case to ease the unfamiliar officers and crews into the highly dynamic business of operating their ships together in close order.[83] The first two days were devoted to moving the fleet to the exercise area in the Bay of Florida. On 3 and 4 February the ships simply got under way together, steamed in columns of vessels abreast by division at double cable length—about 1,500 yards—between ships, and anchored together in the afternoon. On board *Wyoming*, Lieutenant Todd, Lieutenant Costen, Ensign Peck, and Master Day rotated through watches as officer of the deck, keeping station on *Congress* ahead of them.[84] This same work was going on throughout the fleet. Today, with global positioning navigation, computers, and massive arrays of precision instrumentation, formation work is still one of the most nerve-wracking evolutions practiced by modern warships.[85] In 1874, without a pilothouse or any remote instrumentation, it must have been excruciating. One can picture the twelve commanding officers pacing nervously next to their junior officers of the deck, watching as more or fewer revolutions per minute were called down to the engine room to keep the correct distance from the ship immediately in front of them. The rudder was used to follow in the preceding ship's wake. This would have been difficult due to the lack of visibility from the quarterdeck— sailing ships not possessing a raised pilothouse. Modern sailors order changes in the rudder in increments of one-half degree, but the smallest rudder order available to these officers was a "point," or six degrees on the compass. The instrumentation necessary to closely monitor the position of the rudder had not yet been developed. This made formation work that much more difficult.

By 5 February, the formation of columns was executing simultaneous direction changes on signal, as well as the more complicated shift from column formation to line abreast. This was an important fundamental maneuver for steam fleets. Cruising in column abreast by divisions allowed all the vessels in the formation to be within visual range of the flagship and made passing signals easier. The formation for combat, however, was the time-tested line ahead or "column." Transitioning from one to the other would be a vital technique in fleet combat and thus would have been one of the first maneuvers practiced. After a satisfactory day's work, Rear Admiral Case ordered the interval between ships closed to half a cable length (about 350 feet). His confidence in the ships' ability not to run into one another was growing. The fleet executed the reverse of the evolution practiced on 5 February, moving from a simple column to a column abreast by divisions. The recorder noted that the evolution was carried out at a speed of three knots. (The fleet was limited by the best speed of their slowest ship, and more than one of the ships had worn-out boilers that could only get up enough steam to make three knots.) This was a remarkably slow speed for a time when most navies were operating their formations at speeds of up to twelve knots. At that speed the U.S. fleet would not be able to bring a faster enemy to action if they wanted to or run from them if they desired to refuse action. The recorder goes on to note that the movement to column abreast by divisions took one and a half hours to perform.

After closing out 6 February with some flanking movements, the fleet spent the next four days at anchor. On 11 February Rear Admiral Case and Commodore Parker began to step up the pressure on the officers of the fleet. At 11:20 a.m., signal number 63 was ordered: "Divisions from the right of, form columns of vessels, fleet right oblique, right vessels forward." This fairly complicated maneuver was carried out in fifteen minutes. The skill and confidence of the officers of the deck were climbing rapidly. The new tactics had not entirely replaced the old ways, however, as *Wyoming* went to general quarters at 9:45 then (while still responding to signals from *Congress*) exercised at stations for "repel boarders."[86]

After this introduction to fleet operations, the fourteen ships sat at anchor for the next four days. Commander Cushing and the rest of the commanding officers were summoned to the flagship on Saturday to debrief the last three days of steaming in column order. While the commanding officers were gone, there was no rest for the engineers. *Wyoming's* boilers had been a constant source of headache since her commission began, and the constant usage of steam tactical evolutions was beginning to tell on them. Sunday was spent attempting repairs to the port boiler.

The attempted repairs only barely got *Wyoming* through the evolutions on 11 February. On the twelfth Case decided that he had enough and that outside help would be needed to repair *Wyoming's* boilers. Accordingly, the signal "WYOMING EXCEPTED" was made at 6:45 that morning. Soon the fleet weighed anchor without the *Wyoming* and got under way for that day's operations. In the old days of cruising or showing the flag, the order simply would have been given to make sail, and the *Wyoming* would have continued with her business as usual. But there was no room in a steam navy battle formation for a ship that could not get up enough steam to keep station with the rest of the fleet. *Despatch,* a mail steamer, was detailed to take *Wyoming's* place. Two boiler makers from Key West boarded *Wyoming* that afternoon.[87]

Minus the *Wyoming,* the fleet got under way at 10:00 a.m. Settling into the now-familiar column of vessels, the signal was given for the formation to move to "close order," which was one cable length, or 720 feet, apart. With the interval set, the column began to maneuver using simple commands first, followed by increasingly difficult deployments of the battle line. By the end of the day's exercises, the official recorder was able to note that "marked improvement in the execution of evolutions is noticeable throughout the fleet."

Wyoming, the emergency repairs to her boilers complete, was back in the action on Friday. That evening, Case ordered a boat sent on board the flagship with *Wyoming's* signal and order books. Apparently he and Parker were interested in seeing how well the various ships of their fleet were receiving their signals. The weekend was spent at anchor. Once again the commanding officers were summoned to the *Wabash* for a meeting to discuss the weeks' work. Other visiting was going on as well. During the afternoon watch, *Wyoming* sent visiting parties to the *Alaska, Franklin, Colorado,* and *Lancaster.*[88] Officers from the various ships were mingling, exchanging both professional notes and camaraderie in a way that would have been impossible in a cruising navy.

The evolutions began anew on Monday, 16 February. This was a week of increasingly difficult maneuvers. Apparently *Wyoming* had some issues with her speed on Monday, because prior to Tuesday's evolutions, the admiral signaled "Are you in condition to take position in the fleet." Commander Cushing, who no doubt was frustrated with the performance of his boilers, simply replied, "Yes."[89] The week's work culminated in firing exercises on Friday, the twentieth of February. Case was disappointed in the results of the gunnery. According to his report, hits were made only by the *Wabash, Colorado,* and *Kansas.* One again has to imagine Cushing's frustration as *Wyoming* fired her 11-inch gun four times and her 20-pound rifle twice without registering any hits on the target.[90]

At the beginning of the third week of exercises, *Wyoming* was detached to coal. She returned Monday afternoon and anchored, while the rest of the fleet conducted another gunnery exercise. Torpedo exercises were carried out on Wednesday. Case took a personal interest in this evolution, having been the previous chief of the Bureau of Ordnance. Here was an opportunity to see, in practice, many of the innovations his office had worked on and advocated over the past three years. In spite of Case's efforts, the technology of the spar torpedo had not progressed much beyond the weapon that Cushing had employed against the *Albermarle* a decade earlier. Much depended upon being able to use the ship to place the spar in just the right position against the target and then activate the torpedo by means of wires running along the spar. In nineteen attempts, eight failed on the first try, mostly due either to carrying away the spar or damaging the wires in the course of the attack. Eventually, all the participating ships, with the exception of *Alaska,* were able to explode their torpedoes after one or two attempts.[91] Although Case generally referred to these exercises as a success, he could not have felt that the chances of a U.S. Navy ship being able to carry out a torpedo attack in combat conditions were likely. He made it obvious in his reports that much work was needed by the Torpedo School at Newport, Rhode Island, if this weapon was to have any significance as an effective weapon.[92]

That Friday the fleet steamed in double echelon formation at night as they returned to the Dry Tortugas. It was a fitting display of the tactical prowess gained by the officers and crews of the North Atlantic Fleet in their short three weeks of exercises. On 6 March it was the turn of the monitors. Under the direction of Commodore Parker, the *Manhattan, Ajax, Mahopac,* and *Saugus* were taken out for a single day of formation maneuvering. The exercise showed the complete inability of the monitor-type vessels to fight a fleet action. Restricted to a speed of four knots to accommodate the slowest vessel, *Manhattan,* the four ships tried a series of flanking and echelon movements. Parker noted that *Mahopac,* in particular, took nine minutes to turn eight points to port.[93] It was apparent that the idea of engaging an enemy fleet with these vessels, especially when it was probable that an enemy would attempt to ram them, was far-fetched.

Rear Admiral Case's farewell order to the fleet sounded a note of pleasure at the outcome of the exercises, praising Commodore Parker and the efficient execution of tactics under steam by the vessels of the fleet.[94] However, in reality Case's words can only be described as putting the best face on the occasion for the benefit of his subordinates. Given the disappointing results of the gunnery exercises, the torpedo exercises, and the pathetic four-knot attempt to maneuver the monitor vessels, the best that can be said is that the assembled vessels proved the ability of their officers and crews to execute complicated formations with some degree of precision. Their ability to inflict any damage on an enemy formation was doubtful. Case went on to note that "as the practice and effect of exercises in naval tactics, gunnery, and torpedoes is of more importance to the officers who are to *command our future fleets* [emphasis added] than those who are just passing out, I desire to call the attention of the officers to them."[95] Case's reference to "fleets" shows that he felt that the exercises he had just overseen represented the future employment pattern of the North Atlantic Squadron.

Commodore Parker, in not quite as delicate a position as the flag officer commanding the exercises, had somewhat less generous remarks. "The Commodore said that it certainly demonstrated the lamentable condition of the American Navy," reported the *Army and Navy Journal* on 7 March 1874. "The evolutions had thoroughly demonstrated the necessity of an immediate and radical change in the character of the vessels composing the navy."[96] The fleet of the future, according to Parker, would consist of "rams, artillery vessels, and torpedo boats," an argument he made again in a subsequent *Proceedings* article later in the year.[97]

Return to Station Cruising, 1874–1876

In the short term, it was back to business as usual. The maneuvers notwithstanding, the U.S. Navy of 1874 was still devoted to carrying out detached, single-ship operations. After the fleet broke up, the North Atlantic Station returned to its practice of supplying ships singly or in pairs to represent the interests of the United Sates throughout the Western Hemisphere. *Wyoming* returned to the Washington Navy Yard, where she was put out of service on 30 April. The ship's commission was over, and tragically, her captain's life would be over within the year as well. William Cushing was probably suffering from prostate cancer as the exercises were under way, and his condition worsened through the summer of 1874. Taken off sea duty and assigned to the Washington Navy Yard, he died in December. Cushing's final year symbolized the massive institutional change under way in the Navy, as he shifted from operating independently to spending three weeks executing difficult maneuvers in company with eleven other ships, following orders transmitted by a rear admiral. Over the next decade, the Navy Department and the North Atlantic Squadron struggled to find the right balance between showing the flag in foreign ports and maintaining the capability to operate as a combat unit.

After issuing his farewell, Rear Admiral Case took *Franklin, Congress, Alaska,* and *Juniata* and returned across the Atlantic to reestablish the European Station.[98] This left the North Atlantic Squadron with a total of twelve warships, which remained under the command of Rear Admiral Scott for another two and a half months, until he was relieved by Rear Adm. J. R. M Mullany in June 1874.[99] A Civil War veteran who had been at Mobile Bay with Farragut, Mullany had had previous duty as a flag officer commanding the European Squadron as well as the navy yard at Philadelphia. After assuming command of the station, Rear Admiral Mullany was ordered to place his monitors in ordinary in Pensacola.[100] The cruising vessels quickly deployed throughout the North Atlantic Squadron's area of operations. *Colorado* visited Havana, Cuba, briefly in the summer before returning to Key West to act as Mullany's temporary flagship. *Brooklyn* departed in April for a cruise throughout the Windward Islands. She touched at various ports in the Caribbean before returning to Key West in June. *Worcester* had been Rear Admiral Scott's flagship, which put her in an interesting position. After the decision by the Navy Department to turn the combined forces of the North Atlantic, South Atlantic, and European squadrons over to the European Squadron commander, Rear Admiral Case, *Worcester* was tasked with what essentially amounted to a "special mission" to get Scott out of Key West. After Scott turned over command to Case on 3 January, he and *Worcester* departed on an extended deployment to visit Cuba and the Windward Islands, reporting on conditions there in the aftermath of the *Virginius* affair. Discretely returning to Key West on 1 April (after Case departed), she continued as the flagship of the station after Scott, who had already asked to retire, turned over command to Rear Admiral Mullany in June.

Powhatan, as one of the older ships on the station, remained in home waters throughout 1874. After spending some time off the coast of New Orleans, she proceeded to Norfolk, where she was fitted out for special duty—carrying the next commander in chief of the European Station to Portugal in 1875. *Canidaigua* cruised in the Greater Antilles and Virgin Islands throughout the summer of 1874. *Ossipee* left Key West in April, cruising up and down the eastern coast of Central America, and *Wachusett* cruised to various ports throughout the Caribbean in the spring and summer of 1874, before receiving orders to proceed to Norfolk for decommissioning in November. *Kansas* and *Shawmut* both spent the summer carrying out coastal surveys around various parts of the Caribbean.

The monitors remained in port—*Cannonicus* in New Orleans and *Dictator* at Pensacola. Unlike the cruising vessels, these were not designed for long-distance work but to provide local defense against hostile warships. They were armored, with very little freeboard, and carried their main armament in revolving turrets. Their restricted hull volume offered very little living room for their crews, and it was assumed that the officers, at least, would spend a great deal of the time ashore rather than on board.[101] While their nominal fighting power was much greater than the wooden cruisers', the monitors did have a number of significant weaknesses: They were unwieldy and very difficult to maneuver, their top speed was such that any adversary would be free to pick the terms of the engagement, and as far as endurance on the open ocean, they were considered so unseaworthy and their engines so unreliable and inefficient that they usually had to be towed, or at least escorted, by a cruising ship whenever it was necessary for them to move from one port to another. The monitors (or "ironclads," as they were almost universally referred to in the correspondence of the time) were kept partially manned in ports such as New Orleans, Pensacola, and Norfolk, with the understanding that they could be quickly provided with crews in the event of a national emergency.

All of this is to show that, after the fleet exercises in January and February 1874, the forces of the North Atlantic Squadron ceased to be an operating fighting unit. They dispersed, not to operate as a tactical unit for another two and a half years.[102] In part the reason for this was the desire of the Navy Department to go about the normal missions of the Navy—namely, to look after commercial interests. Another reason, however, was the fact that the U.S. Navy did not possess an infrastructure that allowed its ships to spend more than a few weeks concentrated in one location. Until that changed, it was not possible for the forces of the North Atlantic Station to train together as a combat fleet on a regular basis. Despite Rear Admiral Mullany's best intentions, the North Atlantic Squadron did not perform any maneuvers in 1875.

When Rear Adm. William E. LeRoy took command in January 1876, the Navy Department seemed determined to build on the "success" of the 1874 exercises with another round of squadron drills. Moreover, it appears that Mullany's concerns about keeping a large concentration of ships in the Deep South had registered at the department. Upon assuming command, LeRoy was ordered to concentrate his forces at Port Royal, South Carolina.[103] The importance of Port Royal to operations on the eastern seaboard of the United States had been clear since the Civil War. One of the earliest decisions of the Union's Blockade Strategy Board was to seize a likely port on the Carolina coast to facilitate resupply and coaling of warships on blockade duty. By 1866 a large supply operation, complete with a dock constructed by the Navy, was in place at Port Royal, South Carolina.[104] After the war the facilities at Port Royal were ordered closed, but the Navy Department elected to retain ownership of the property.[105] When the threat of yellow fever rendered the navy yard at Pensacola inhospitable, the idea of using Port Royal as a rendezvous was resurrected. Here the ships could be serviced without incurring the expense of traveling north to the next-closest shipyard at Norfolk, 460 miles away. An additional benefit was that the warships would remain under the operational control of the commander in chief of the station, not the commandant of a navy yard. Port Royal had the additional advantage of being relatively close to the action in the Caribbean, especially when compared with Hampton Roads.[106] In 1876 the monitors that previously had been in ordinary at Pensacola, Florida, were also moved to Port Royal, where they could be anchored in fresh water, which was preferable for the preservation of the monitors' metal hulls.[107] The USS *New Hampshire*, an outdated ship-of-the-line, was ordered to Port Royal as a permanent depot ship. The entire operation was placed under the command of Capt. J. B. Clitz, who was given the title of commodore and the responsibility for all the warships at the Port Royal rendezvous when the commander in chief was not present.[108] This last stipulation is crucial to understanding how this was a move by the Navy Department toward an operational cycle for the squadron that would involve the vessels spending more time in close proximity. One way to facilitate this was to create a "rendezvous" where the ships could gather that did not create chain-of-command difficulties with the commandant of a navy yard. This seems to indicate that the Navy Department in 1877 was moving away from a cruising mindset to one that involved a concentration of its forces. Port Royal continued to grow in importance in the eyes of the Navy Department and the North Atlantic Squadron. Opening as a "naval depot" in 1876, by 1877 it was classified as a "naval station" and had its own commanding officer.[109]

The department was eager to try out its new home port in 1876 and sent a lengthy order to Rear Admiral LeRoy upon his assumption of command of the North Atlantic Station. After a preliminary period in which the squadron would practice drills and gunnery exercises while at anchor or in the vicinity of Port Royal, LeRoy was expected to get under way and lead his squadron on a deployment to the Caribbean, exercising them along the way in "squadron evolutions and tactics." "The Department hopes," wrote Secretary Robeson, "that with the class and conditions of vessels in this command, and with the preliminary drill, the squadron will be able to make a creditable exhibition of our Naval power, wherever it may go and whatever work it may be directed to perform."[110] Soon after taking command, LeRoy issued General Order No. 6, which organized the "vessels attached to the North Atlantic Station." They were divided into three divisions, with a reserve division composed of support ships and monitors.[111] In the beginning at least, LeRoy was thinking in terms of commanding his forces as a single combat unit, and there was an expressed desire both on his part, and that of the department, to project a squadron into the Caribbean in support of U.S. foreign policy objectives.

Unfortunately, real-world contingencies intruded upon the plans Admiral LeRoy and the department had for the vessels of the North Atlantic Station, and as will be seen, the prevailing model of sending individual ships abroad to "put out fires" continued. Things began to unravel in March, when unrest in Haiti threatened U.S. interests. On 14 March a terse telegram from the Department of the Navy ordered LeRoy to dispatch two vessels to Port-au-Prince without delay.[112] Unrest in Mexico was on the administration's mind as well, and two days later, on 16 March, LeRoy was ordered to take four ships, *Hartford, Swatara, Shawmut,* and *Marion,* to Tampico, Mexico.[113] While these orders, and subsequent ones concerning these four ships, refer to this as a "squadron," the plan for them upon arrival in Mexican waters was to "distribute" them as most advisable, making sure that at least one ship was stationed at the mouth of the Rio Grande del Norte. On 1 April the U.S. government ordered LeRoy to have one of his other ships visit ports along Mexico's eastern coast.[114] With no chance remaining for the ships to conduct exercises, *Marion* was ordered to the Mediterranean on 17 May, and Rear Admiral LeRoy was ordered to bring *Hartford* north to Philadelphia on 24 June.[115]

While the contingent events intruded upon the department's extensive plans to hold squadron maneuvers in the summer, the North Atlantic Squadron was eventually able to hold a much-scaled back version in the fall, although LeRoy was no longer in command.[116] Rear Adm. Stephen Decatur Trenchard assumed command of the station in September 1876. He was immediately ordered to assemble whatever ships he had available in Port Royal in October of that year for an inspection and exercises in naval drill prior to departing to their cruising stations for the winter.[117] After all the attention paid to squadron exercises in steam tactics, drills with torpedoes, and the like in LeRoy's original orders of 28 January, the centerpiece of the October maneuvers, or rather the only aspect the secretary of the navy felt was worthy of his including in his Annual Report to Congress for 1876, was the landing of a naval brigade—a fighting unit consisting of the officers and men of the assembled ships.[118] Although we can see early evidence that the department was becoming more interested in the ships of the North Atlantic Squadron spending meaningful time together as a combat unit, the secretary's attention is still drawn to the Navy's ability to project combat power ashore, providing security for U.S. business interests on land. In an era when the U.S. Marine Corps still mainly consisted of small detachments to keep order on board ships, it was expected that a commanding officer could deploy a fighting unit from the men of the entire ships' company.[119]

The twelve vessels under the command of Rear Adm. William E. LeRoy were never referred to as a "squadron." Indeed, LeRoy did not identify himself as a "squadron commander," nor is there any evidence that, other than issuing General Order No. 6 assigning his ships to divisions, he had any intention of commanding the assembled vessels as a squadron or fleet. LeRoy's job as commander in chief consisted largely of managing the logistics involved with the individual movements of his ships.

The years 1877 and 1878 were lean ones for the North Atlantic Station. Many of the wooden cruising vessels that had carried out the large exercises of 1874 and the smaller reprise of 1876 had been placed out of commission. In 1877 Rear Admiral Trenchard had five cruising vessels to work with. These five were sent on deployments that touched down the eastern seaboard and into the Caribbean. This was still the overarching mission of the U.S. Navy, as the secretary of the navy made clear in his remarks in his annual report to Congress.[120] Protecting commerce and commercial opportunities was, however, not just a mission for overseas.

The year 1877 brought an opportunity to exercise the naval brigade concept in a way that the Navy Department had probably not conceived of using it. During the Railroad Strike of 1877, sailors from the North Atlantic Squadron provided security for government buildings and facilities in Washington, D.C., including the Washington Navy Yard.[121]

By 1878 the five vessels Trenchard had commanded in 1877 had been whittled to two: *Powhatan* and *Plymouth*. There is no evidence that these ships either spent any sort of significant time together in port or cruised together.[122] Squadron tactics under steam were, for the moment, no more than a memory. Rear Admiral Trenchard gave up his command in September 1878, relieved by Rear Adm. J. C. Howell.[123]

Howell had been appointed a midshipman from Pennsylvania in 1836. After distinguished service during the Civil War, he served as chief of the Bureau of Yards and Docks from 1875 to 1878, before being assigned as the commander in chief, North Atlantic Squadron. Howell's two-ship squadron became a one-ship squadron after an insurrection broke out on the island of Santa Cruz in the Caribbean. The Navy Department cabled Howell, who was in Portsmouth, New Hampshire, flying his flag on board *Powhatan,* to send a vessel at once.[124] The department was less than satisfied with the North Atlantic Squadron's response. Of Howell's two ships, *Powhatan* was too far away to have any effect on the situation, and *Plymouth,* which was closer but in port, had not provisioned and was unable to get under way. In the event, a French ship got to Santa Cruz first and landed troops, but not before the American consulate had been burned.

It is likely that this event solidified the Navy Department's resistance to keeping ships in northern ports and redoubled its determination to have a facility, such as the one at Port Royal, where a concentration of North Atlantic Station assets could be ready to sail from on a moment's notice.[125] After reprimanding Rear Admiral Howell for not having a warship ready to respond, the department had the nerve to tell him that the "Department will leave to your discretion any further movements of your squadron," as if he had control of more than two ships.[126]

In January Rear Admiral Howell was appointed to command the European Squadron, and Rear Adm. R. H. Wyman took command of the North Atlantic Station.[127] In keeping with tradition, Wyman had performed a previous tour as the chief of a bureau—in his case the Navy Hydrographic Office. During the eight years he held this position, Wyman was responsible for instituting a worldwide program of charting and surveying that decreased U.S. Navy reliance on foreign (often British) sources for their charts.[128]

Rear Admiral Wyman was immediately greeted in his new command by a telegram from the Navy Department informing him that there was a disturbance taking place at Puerto Caballo, Venezuela, and directing him to prepare his flagship to get under way to proceed there at once.[129] At this point the flagship represented the only vessel Wyman had under his control, so that essentially, upon receipt of that telegram, the entire North Atlantic Squadron set out for the Caribbean. Touching at Cuba, *Powhatan* was greeted with another telegram that instructed her to proceed at once to Puerto Caballo and "look after American interests in that quarter."[130] Things cooled down, however, and Wyman was back in Norfolk by April.[131] There was trouble in Panama in June, but the Navy Department decided that sending a vessel was not warranted.[132] By 1879 five ships were assigned to the station, as the *Essex*-class wooden screw sloops began to report for service. There is no evidence, however, that any of these ships carried out either squadron-level tactical exercise or cruising deployment during the year.

The importance of the Central American region was growing as more and more thought began to be given to the project of putting a canal through the isthmus.[133] One of the ironic results of the Railroad Strike of 1877 was that the personal involvement of naval officers, who already felt that their mission largely involved protecting American commercial interests overseas, in the maintenance of civil order convinced them that the way to prevent future unrest was to have more and more commercial opportunities overseas, providing necessary outlets for production surplus and, by extension, jobs and money for the working class.[134]

Just as the question of a canal through the isthmus began to capture the imagination of the American people, the attention of the Navy Department was required in the north as well. The fisheries questions that had been thought settled by the 1871 Treaty of Washington began to be raised again. In late July 1879, *Kearsarge* was sent to Charlottetown, Prince Edward Island, Canada, where she was met by a special agent of the Department of State who had been sent to gather information about possible treaty violations in the fishing waters off Canada.[135] The North Atlantic Squadron was still very much a policing service for the State Department.

Conclusions

The Key West exercises of 1874 signified recognition by the Navy Department that conducting fleet tactics under steam would be a skill required of its officers in the future. The continued development of U.S. strategic interest in the Caribbean meant that it was no longer enough to simply train naval cadets in tactical theory at the academy. The process of developing a fleet capable of fighting in formation could only be accomplished by rigorous exercise at sea. Although managing the individual movements of his ships was still the best description of what the commander in chief did on a daily basis, it was now clear that he would be spending an increasing amount of time leading his vessels in formation. The conflict faced by the North Atlantic Squadron between these two distinct mission types would become evident over the next two decades. While the Royal Navy possessed enough warships, armored and unarmored, to field different squadrons for cruising and fleet operations, the U.S. Navy did not. The North Atlantic Squadron would be forced to attempt to maintain proficiency in both missions with the assets available.

The squadron did not yet have a permanent home port, although the Navy Department had been considering various options for one. Nor did they yet have different classes of vessels that would utilize different weapons systems to fight in coordination with one another against an enemy force, although experiments were being carried out in Newport with the torpedo vessel *Alarm*. Methods of rapidly transmitting tactical signals throughout a formation were a subject of intense focus for the chief signal officer of the Navy. In short, the vessels on the North Atlantic Station could not yet be referred to as constituting a fleet, although the elements of a fleet were beginning to take shape. The 1880s would see two major developments. The authorization of the U.S. Navy's first four steel warships in 1883 would herald the arrival of a "naval renaissance."[136] Two years later, an officer who would almost singlehandedly be responsible for the development of fleet tactical doctrine would set up shop across Newport Harbor from the Torpedo School. Commo. Stephen B. Luce would not only become the first president of the Naval War College, but he would follow that accomplishment with a tour as the commander in chief of the North Atlantic Squadron, where he would put in action many of the tenants he had so fervently preached at the college.

2

Toward a New Identity, 1882–1888

The decade of the 1880s witnessed a number of centennial celebrations across the United States, as various anniversaries associated with the founding of the nation were observed. This was particularly true for cities along the East Coast such as Philadelphia and New York, where celebratory showings by the Navy were popular. It was a time when the legacy of the Revolution was being reconciled with the horrific experience of the Civil War to forge a new identity for the nation. Into this heady atmosphere of celebration and new possibilities came something that had been missing in the 1870s: a revived postwar economy. With the federal government enjoying a budget surplus for the first time since the war, it was a favorable time for a discussion about the nation's future naval policy. These discussions provided the foundation for the naval renaissance that was to come, but as they were being carried out, identity confusion and conflict of missions continued for the North Atlantic Squadron.

Organizational change in the North Atlantic Squadron took place slowly. In the squadron's operations during the 1880s, the impetus for a shift in identity can clearly be identified. Concern about European encroachment in the Western Hemisphere, the rise of South American nations, and a war scare with Chile created the context within which a debate about the Navy's mission and matériel took place. The first ships of the New Steel Navy were authorized in 1883 but would not enter active service until almost six years later. Although matériel and structure did not change significantly in the early to mid-1880s, the mission did. The squadron became concerned not solely with cruising operations but also with training exercises to prepare for combat as a unit. Naval strategists, through the work of two separate naval advisory boards and numerous articles in the *Proceedings* of the U.S. Naval Institute, officially insisted that the U.S. Navy's proper mission was to guard the coastline and raid enemy commerce. However, during this time, the North Atlantic Squadron regularly engaged in fleet maneuvers. By 1888 something such as the Key West exercises of 1874 was, if not routine, at least commonplace.

The theoretical justification for tactical exercises was provided by Rear Adm. Stephen B. Luce. In the years before his groundbreaking work in establishing the Naval War College, he advocated a synthesis of operational exercises and classroom theoretical work and attempted to implement such a program during his tour as commander in chief of the North Atlantic Squadron. But regardless of the desire, either on the part of the Navy Department or the commander in chief, to carry out fleet tactical training, both Rear Admiral Luce and his predecessor, Rear Adm. James E. Jouett, spent most of their time carrying out diplomatic tasking and protecting U.S. commercial interests rather than training their forces for fleet combat. However, exercises did occur. While the establishment was still publicly promoting a cruising navy, the North Atlantic Squadron was repeatedly practicing as if the intended primary mission was to meet an enemy in a fleet engagement at sea.

Policy and Matériel Debates: *Proceedings* and the Naval Advisory Boards

The election of James A. Garfield of Ohio to the presidency in 1880 placed the moderate, or "Half-Breed," wing of the Republican Party in power.[1] The leader of this faction was Maine senator James G. Blaine. As a reward for his efforts on Garfield's behalf, Blaine was appointed secretary of state in the new administration. Blaine was an energetic and outgoing politician who desperately wanted to be president. His vision for U.S. foreign policy included dominance of the Western Hemisphere and increased trade both there and throughout the Pacific. His particular interest in Latin American affairs indirectly encouraged the nascent naval reform movement. In an attempt to head off possible British involvement, he injected the United States into the still-raging War of the Pacific (1879–84), backing Peru against Chile.[2] It was a politically risky move, as Chile had largely established military superiority by this point in the war. Chilean naval power at the time was centered on two central-battery ironclads, *Cochrane* and *Blanco Encalada*. The U.S. Navy of 1881 had nothing to deploy in the Pacific that could contend with either British or Chilean naval assets, a fact of which U.S. officials were uncomfortably aware. As in the 1874 *Virginius* affair, a conflict existed between the foreign policy objectives of the U.S. government in the Western Hemisphere and the naval means of obtaining those objectives. Naval officers in the early 1880s worked to address this conflict. Far from being the mindless advocates of offensive naval power as they are sometimes characterized, the published record shows that officers who led the push for naval reform did their best to reconcile national strategic requirements and operational naval capabilities.[3] The professional forum provided by the U.S. Naval Institute offered an opportunity for these officers to voice their policy recommendations.

For its 1880 prize essay contest, the Naval Institute asked members to write on the subject "Naval Strategy for the United States." The submissions were evaluated by a panel of three distinguished judges: Secretary of State William M. Evarts, Secretary of the Navy Richard W. Thompson, and Senator John R. McPherson, chairman of the Naval Affairs Committee. The winning entry, submitted under the motto *Sat cito, si sat bene,*[4] was written by a lieutenant who was a member of the Naval Academy faculty and the institute's secretary: Charles Belknap. Belknap reviewed the reasons for maintaining a strong Navy: "the unsettled condition of society in the less civilized parts of the world; the depressed state of our maritime interests; the enforcement of the principles of the Monroe Doctrine and of our neutral rights."[5] He then translated those reasons into four operational requirements: "the naval defense of our coasts and sea ports . . . the protection of our commerce and the destruction of an enemy's . . . the destruction or capture of the men-of-war of an enemy [and] . . . carrying the war into an enemy's country."[6] Although the stated order of the operational requirements seemed to privilege the historic priorities of coastal defense and commerce raiding, it is clear that Belknap was calling for the development of an offensive capability. Belknap reminded readers of the diplomatic "draw" between Spain and the United States that ended the *Virginius* affair and warned against "our again being placed in such a false position before the eyes of the world."[7] Belknap's essay may be taken to represent the thinking of junior officers in the 1879–1880 period. The fact that the essay was judged ahead of seven others by the three men most influential in naval matters in the Hayes administration is suggestive that there was some consensus as early as 1879 that the United States Navy should possess the capability in certain circumstances to conduct fleet operations against an enemy.

The specifics, however, of the matériel necessary to carry out this function were not so easily agreed upon. The 1881 prize essay attempted to address this problem by posing the question of the "type of (I) armored vessel, (II) cruiser, best suited to the present needs of the United States." The winning essay was submitted by Lt. Edward W. Very from the Navy Signal Office. If any junior officer in the Navy was intimately involved with the real operational questions of close order formation, it was Lieutenant Very, whose work in perfecting a night signaling system for warships in formation will be discussed below. Yet his prize essay gave a very conservative answer to the issue raised. In it Very recommended a two-pronged construction plan. Improved monitors for coastal defense would be aided by cruising vessels with full sail rigs to attack enemy shipping and protect U.S. overseas commercial interests. Nowhere in his essay is the requirement to operate in close order with other warships mentioned. The honorable mention essay, written by Lt. Seaton Schroeder of the Hydrographic Office, differed only slightly from Very's in the recommendation for the type of armored vessel to be used for coastal defense. Schroeder favored a version of what the Europeans were calling a "central-battery ironclad," rather than a monitor.[8]

This inconsistency, then, between policy ambition and matériel reality was the major challenge facing the Navy in the 1880s. Lieutenant Belknap claimed that the capability to engage enemy combatants on the high seas was necessary. However, the next year, neither Lieutenant Very nor Lieutenant Schroeder spent much time on this dimension of naval warfare in their essays, discussing instead the relative merits of monitors and sail-rigged cruisers. None of the three essays discussed fleet maneuvers, fleet combat, or the ability of the recommended types of armored or cruising vessels to be built to be able to operate together in close order. Historian Peter Karsten accuses naval writers who published essays during this time of artificially keeping their work conservative, thereby making it easier for Congress to vote them money. However, a close reading of Lieutenant Very's other correspondence throughout his active duty career renders it unlikely that he, as a lieutenant, cared what the establishment thought.[9]

With the Republican ascent to power in the 1880 election, President Garfield's new secretary of the navy intended to do something about this identified inconsistency between stated mission and anticipated wartime operational employment of the Navy. William H. Hunt was a former Confederate officer and attorney general of the state of Louisiana. Although his tenure as secretary lasted only a little over a year, Hunt understood that a major reason for the congressional reluctance to spend more for the Navy was that, as the prize essays for 1880 and 1881 showed, there was no clear consensus on either what the national naval strategy should be or what kind of warships should be built. To remedy this, Hunt appointed the Naval Advisory Board, under the leadership of Rear Adm. John Rodgers (of the well-connected Rodgers naval family), to prepare a single, coordinated recommendation to Congress about the future of the Navy and its warships. The board was instructed to meet on 11 July 1881 and to transmit a report by 10 November.[10] It consisted of fourteen officers, representing the line as well as engineers and naval constructors. After meeting for the four months the secretary had allowed them, the board's members found that they were unable to reach a consensus. They submitted two reports: a majority report, signed by eleven members and largely representing the views of the line officers, and a minority report, signed by Chief Engineer Isherwood and three other engineers and naval constructors.

The introduction to the majority report stated, "At present the unarmored vessels of the service are the only ones required to carry on the work of the navy."[11] The line officers were willing to leave it at that. They acknowledged that armored warships would be required in time of war but expressed confidence that the national production capabilities would be enough to produce these ships on demand.[12] The report went on to review the vessels available to cruise on foreign stations against the number they felt was sufficient, including a reserve. The result was a recommendation for the construction of thirty-eight new unarmored cruising vessels. Two would be capable of sustaining a speed of fifteen knots, six would be designed to maintain fourteen knots, ten could attain thirteen knots, and twenty would have a maximum speed of ten knots. The ten-knot vessels would be of wooden construction, the rest were to be built of steel. All of the vessels would have full sail power as well as steam engines. The majority supplemented their building plan with a call for smaller ram vessels as well as torpedo vessels—to be used in conjunction with the existing monitors for harbor defense.[13]

The minority report, representing the views of Chief Engineer Isherwood and the Bureau of Steam Engineering and Bureau of Construction and Repair, agreed with the majority that the current needs of the Navy did not include the ability to engage armored warships in combat. They were, however, more forceful about pointing out that such a capability was necessary to be considered a first-rate navy.[14] The minority report pointed out that unarmored cruising vessels "enable[d] a naval organization to be maintained by serving as training vessels for crews and officers during peace."[15] The report went on to reject entirely the construction of new wooden vessels and instead insisted that all vessels should be of iron. This represented the pragmatic view of the engineers and naval constructors that the U.S. steel industry was not capable of producing the amount and quality of steel necessary to build a ship. Overall, the minority recommended the construction of two spar-deck ships (frigates), six first-rate single-deck sloops, ten second-rate sloops, twenty gunboats, and fifteen steam torpedo boats. All were to be constructed of iron. After reviewing the documents, the full House Naval Affairs Committee eventually recommended the construction of a total of fifteen ships and reported the bill to the floor of Congress for a vote. After the bill went to conference, only two ships were finally authorized, but no funding was appropriated for their construction. Congress expected the Navy Department to build the new warships with construction and repair funds that had already been appropriated. A small beginning, but it was acknowledgment nonetheless that new, nonwooden warships were considered necessary by Congress.

The assassination of President Garfield in 1881 and the subsequent elevation of Chester Arthur to the presidency led to a change in the secretary of the navy's office. William H. Hunt, who was ill and having difficulty carrying out his duties as secretary anyway, was posted to St. Petersburg as the U.S. minister to Russia.[16] This cleared the way for President Arthur to repay some political favors.

Arthur's appointee to head the Navy Department was William E. Chandler.[17] A New Hampshire state assemblyman and newspaper editor, Chandler turned out to be a surprisingly effective advocate of increased naval spending and a new construction program. He appointed the second Naval Advisory Board on 5 August 1882 to make recommendations concerning the construction of the two cruisers authorized by Congress earlier that year. This board was chaired by Rear Adm. Robert W. Shufeldt and consisted of six other members, including a civilian naval architect and a civilian marine engineer. Shufeldt's board submitted its recommendations on 21 November 1882. Having been witness to the reception on Capitol Hill of the first Naval Advisory Board's report, the members of this second board were somewhat more pragmatic. In addition to the two warships already authorized by Congress, they recommended two more second-rate single-deck unarmored cruisers for a total of four ships, all to be constructed of steel. The New Steel Navy was born on 3 March 1883, when Congress appropriated $1.3 million for the construction of the four ships, to be named *Atlanta, Boston, Chicago,* and *Dolphin*—the so-called ABCD ships.[18]

The significance of the reports of the two advisory boards and the eventual congressional construction approval is to point out the continued conflict between matériel and the professional expectations placed on the North Atlantic Squadron. The ABCD ships were cruising vessels, unarmored and with full sail rigs. However, as the decade progressed, the North Atlantic Squadron was increasingly expected to maintain proficiency at fleet tactical drills and close order steaming. As Secretary Chandler put it in his 1883 remarks, "It is not now, and it never has been, a part of that policy to maintain a fleet able at any time to cope on equal terms with the foremost European armaments. . . . We unquestionably need vessels in such numbers as fully 'to keep alive the knowledge of war,' and of such a kind that it shall be a knowledge of modern war; capable on brief notice of being expanded into invincible squadrons."[19] While Congress was debating the missions and composition of the new Navy, the North Atlantic Squadron rendezvoused in Port Royal in April 1882 and steamed north to Hampton Roads for a change of command and exercises in fleet tactics off Fort Monroe.

Squadron Exercises, 1882

On 1 May 1882, Rear Admiral Wyman was relieved by Rear Adm. George H. Cooper as commander in chief.[20] Cooper was a native of New York and was the last nineteenth-century commander in chief of the North Atlantic Squadron not to have attended the Naval Academy at Annapolis. He joined the Navy in 1837 as a sixteen-year-old midshipman and spent four years in the *Constitution* before being sent to the naval school at Philadelphia to prepare for his exams.[21] Service in the Mexican and Civil Wars followed. He had been the commanding officer of Rear Adm. John Rodgers' flagship, *Colorado,* in the Asiatic Fleet during the Korean Expedition of 1871, where he gained experience conducting squadron-level operations. Subsequently he had commanded the navy yard at Pensacola, Florida, before his assignment as commandant of the New York Navy Yard.[22] The change of command took place in Hampton Roads, Virginia, where the squadron had been ordered to assemble.[23] All six warships of the North Atlantic Squadron were in attendance: *Tennessee, Kearsarge, Vandalia, Alliance, Enterprise,* and *Yantic.*

Launched in 1865, the flagship *Tennessee* represented the culmination of the technology marrying wooden sailing ships with steam auxiliary power. At a displacement of 3,200 tons, she was large and roomy and was a coveted assignment for sea duty.[24] *Kearsarge* was the most famous and decorated of the Navy's Civil War–era steam sloops, being the celebrated veteran of the epic battle with the Confederate raider *Alabama.* The less famous screw sloops *Vandalia* and *Enterprise* and the gunboats *Alliance* and *Yantic* rounded out the squadron.

The Navy Department was anxious to take advantage of the rare opportunity of having the ships concentrated to conduct fleet tactical exercises. Much of the department's newfound drive to carry out these exercises probably had to do with the energetic new chief of the Bureau of Navigation and Detail, Commo. John G. Walker.[25] Within two days of assuming command, the department (probably Walker) had cabled Cooper with instructions to take his squadron to sea for exercise at steam tactics. Cooper promised to get to sea by 10 May.[26] There is some evidence that the order to conduct exercises initially caught Cooper off guard. His correspondence during the eventual twenty-day at-sea period makes several oblique references to the speed at which his commanding officers had prepared their ships for sea and commends them for being able to stay at sea for so long on such short notice.[27] The printing press on *Tennessee* was kept busy as a flurry of general orders and circular instructions to commanding officers were quickly produced, outlining Rear Admiral Cooper's organization of his squadron.[28]

On 10 May 1882 Cooper's six ships got under way and stood out from Hampton Roads. They spent the next two days scattered in heavy fog but eventually were able to commence steam maneuvers on 13 May. With a slight northwest breeze, smooth seas, and his ships making an average speed of 4.5 knots under steam, Cooper ordered the squadron to form in simple echelon at "open order." This allowed commanding officers to maintain position with a greater interval between ships than that called for in the tactical manual. Cooper explained in his after-action report that "close order" forced the expenditure of larger amounts of coal, as commanding officers had to constantly use higher engine settings to achieve and maintain their positions. The supply of coal was on Rear Admiral Cooper's mind throughout these exercises.[29]

Cooper began with the same basics Rear Admiral Case had practiced eight years before. For any squadron commander, the fundamental formation skill was the ability to move his ships from line to column and back. A line abreast formation enabled a commander to spread his ships out in such a manner that they could maximize the amount of ocean searched for opposing forces yet remain in visual contact with the flagship. The column allowed him to concentrate his firepower at the onset of battle. The ability to rapidly shift from line to column could mean the difference between victory and defeat in a fleet engagement.[30] Over the course of the next two days, the maneuvers became more difficult, until the tactical exercises were completed on 15 May. The squadron spent that afternoon conducting a number of turning trials to, in Cooper's words, "ascertain the relative handiness of the different vessels in turning."[31] His written orders to each of his commanding officers stipulated that measurements would be taken at full speed and two-thirds speed, with the helm half over and hard over. The idea of spending extensive amounts of time in formation was still new enough that accurate data on the turning abilities of U.S. ships did not exist. In giving his instructions for measuring tactical diameter, Admiral Cooper referenced a work by Chief Constructor W. H. White of the Royal Navy titled *Turning Powers of Ships*.[32]

The remainder of the time at sea was spent under sail. The secretary of the navy's report for 1882 makes special mention of the twenty days the ships of the North Atlantic Station spent cruising "in squadron," but in reality, steam maneuvers only took place from 13 to 15 May. The remainder of the squadron's time under way was spent under sail. Rear Admiral Cooper was adamant about keeping the ships' crews busy with sail and spar drills and other training on board the individual ships. These drills were much more in line with the daily operations of a navy whose chief mission was cruising and "showing the flag." Target practice with the ships' guns was carried out on 20 May, and torpedo practice took place on 24 May. The squadron returned to Hampton Roads on 30 May 1882.[33] In twenty days Rear Admiral Cooper had already spent more time in direct tactical control of his warships than the previous five commanders in chief combined.[34]

Rear Admiral Cooper and the Limits of Wooden Cruising Vessels

Upon his return to Hampton Roads, Rear Admiral Cooper was pleased to learn that the Navy Department intended to keep the squadron together during the summer. This would afford many more opportunities for training in fleet tactics under steam.[35] An indication that Cooper intended to spend the summer working was his 2 June circular letter to commanding officers in which he actively discouraged officers from applying to take leave for periods longer than twenty-four hours and required that the papers be submitted to him for approval if any did.[36] After routinely dealing with yellow fever year after year, the Navy Department suspended the practice of sending warships to the Caribbean during the hot fever season and replaced it with the practice of moving north in the summer for exercises then dispersing throughout the Caribbean as the weather cooled and the threat of yellow fever decreased.[37] Adm. David Dixon Porter endorsed this new convention, noting in his annual report to the secretary of the navy that "cruising together the past summer has been of great advantage to the squadron in many respects, and I recommend that the practice be kept up."[38]

After a couple of weeks to perform minor repairs and resupply the ships, *Tennessee, Vandalia, Alliance,* and *Yantic* got under way on 17 June. *Kearsarge* and *Enterprise* needed more substantive repairs and so stayed behind at the Norfolk Navy Yard.[39] Once under way, the squadron immediately began to do formation work, as ordered by the Navy Department.[40] On Sunday, 18 June, after divine services, the officer of the deck noted that "the Chief of Staff exercised the fleet in naval tactics under steam."[41] He went on to note in his entry that he had to "revolve the engines by the bell during the maneuvers." The methods of rapidly sending engine orders from the deck to the engine room were still being worked out. This set of exercises began more aggressively than the previous one, no doubt owing to the experience that the ships and their crews had received the previous month. In close order at seven knots, the four ships shifted first into echelon, then line abreast, then formed two columns of two in a respectable eight and a half minutes. The evolutions continued over the next two hours, the squadron working through line abreast, column, and echelon formations before maneuvering ceased for the day.[42] The four ships followed the same routine for the next four days, 19–21 June, spending at least two hours each day maneuvering in response to tactical signals from the flagship. On 21 June each ship performed more maneuvers to test and chart its tactical diameter.[43] Once the fleet exercises were complete, *Vandalia* was detached and sent to Portsmouth, New Hampshire, while *Alliance* was sent to Boston. In his after-action report, Rear Admiral Cooper noted that, in accordance with his instructions from the Navy Department, "all the evolutions laid down in Parker's Steam Fleet Tactics were made that were possible with a Squadron of four vessels. . . . The vessels were maneuvered in closer order than during the previous cruise, and more care was observed in preserving proper positions." He went on to offer his recommendation that "in future, the time to be devoted to these exercises be shortened."[44] Cooper's correspondence throughout 1882 gives the impression that he found tactical exercises useful to a point, but he was not as excited about them as Rear Adm. Stephen B. Luce or Rear Adm. John G. Walker would be in a few years. Cooper carried out tactical exercises with his wooden cruising vessels not because he was preparing to fight a fleet action but because he thought formation work in appropriate doses was good professional development for his officers and because the Navy Department repeatedly directed him to throughout the summer.

Tennessee and *Yantic* arrived in New York City on 22 June. It was a homecoming of sorts for Rear Admiral Cooper, as he had been the commandant of the New York Navy Yard before his assignment as the North Atlantic Squadron commander in chief. They stayed at anchor there for the next twenty days, taking on coal and supplies from the New York Navy Yard. On 12 July, *Tennessee* and *Enterprise* set out for Boston by way of Provincetown, Massachusetts, where they were to meet up with *Yantic* and the other vessels of the squadron.[45] The Navy Department continued to press Cooper to conduct squadron exercises at every opportunity.[46] On the way to Boston, *Tennessee, Enterprise, Alliance,* and *Yantic* carried out exercises in steam fleet tactics on 31 July. The after-action report is unremarkable. All the usual combinations of column, line abreast, and echelon formations were practiced. This time the base speed was eight knots and everything was done at close order. At the conclusion of the fleet drills, fires were banked and the remainder of the underway period was spent under canvas. The crews were worked at spar and sail drills, as well as general quarters, both day and night. On the fourth day under way, steam was raised in order to hold target practice. The four ships moved in a circle around the targets at ranges from 1,000 to 2,500 yards. Cooper was pleased with the results, noting that Seaman N. P. Peterson of *Tennessee*'s No. 12 gun (an 8-inch rifle) had struck the target at eight hundred yards.[47] It speaks to the low state of efficiency of naval artillery in the 1880s, prior to range-finding and sighting equipment, that the attainment of a single hit was cause for the commander in chief to mention the gun captain by name in his report.

Overall, Cooper was pleased with his ships' performance on the trip to Boston. He reported that the enthusiasm shown by the officers and men "goes to prove the utility of squadron exercises, as long as circumstances will admit."[48] He went on to say, "In carrying out the views of the Department in this matter, I make it an object to keep every one on the alert. . . . At the same time I am very careful not to worry or harass the command with anything like overwork."[49] Cooper represents the epitome of the commander in chief in transition. He recognized the utility of squadron exercises and carried them out professionally (when directed to by the Navy Department). However, without fail, he ordered his cruising vessels to bank fires and spread canvas at the first available opportunity. With the matériel in place on the North Atlantic Station in 1882, sail was still the primary method of propulsion.

The squadron anchored in Boston on 2 August 1882.[50] After three days in Boston, the four ships headed to Portsmouth, New Hampshire, and Portland, Maine.[51] Cooper held brief tactical exercises as the squadron departed the harbor under steam power, but once out to sea, fires were banked and the squadron continued under sail alone. Exercises in wearing and tacking, making and shortening sail, and reefing and shifting topsails were carried out.[52] At night the squadron exercised with Very signals, a new method of night communication using flares that would eventually replace the not-very-reliable Costen lights.[53] The signal office had been hard at work on perfecting a system to maneuver large numbers of ships at night and after testing several systems had settled on the rockets designed by Naval Institute essay-winning Lt. E. W. Very. Testing would continue over the next few years, but little was found that enabled quicker or easier understood night signaling than the Very system.[54] The interest and amount of effort expended by the signal office to tackle the night signaling problem is yet another indicator of the operational Navy's interest in being able to sail and fight in tactical formations.

At Portsmouth the squadron was reviewed by President Chester A. Arthur and Secretary of the Navy William Chandler, the two dignitaries having arrived at Portsmouth in *Despatch* on 9 September. Over a period of three days, official visits were exchanged. President Arthur and Secretary Chandler visited *Tennessee* on 11 September and were able to witness tactical drills as well as target practice. Arthur, who had just assumed his office following the death of President Garfield, would prove to be a great friend of the Navy. He and Chandler were appropriately pleased with the state of training of their naval forces. Rear Admiral Cooper was as well, noting that "in performing evolutions under steam in close order, the commanding officers showed much skill and confidence in handling their vessels." Cooper was genuinely concerned with training for his junior officers. He went out of his way to ensure that each of them was given ample opportunity to act as the officer of the deck during tactical maneuvers, responsible for directing his ship's movements with proper rudder and engine orders. North Atlantic Squadron General Order No. 14, published on 8 September, mandated that after-action reports from fleet tactical exercises would list each line officer and give the times each had stood watch as officer of the deck during maneuvers. Cooper insisted that ensigns and midshipmen get more time as the officer of the deck to enhance their professional development.[55]

The presidential review signaled the end of tactical training for the squadron in 1882. One afternoon was devoted to fleet tactics on the way to Portland, Maine, from Portsmouth—a training period insignificant enough that it did not warrant an after-action report from Cooper. It was time for the squadron to break up and send the individual ships on their way with their cruising assignments in the West Indies. After a squadron visit to Philadelphia in October, the warships moved south to Hampton Roads, Virginia, where they departed for the winter cruise on 12 December 1882.[56]

The flagship operated alone on the winter cruise. *Tennessee* traveled first to Martinique, then St. Christopher's Island, St. Thomas, Santa Cruz, and, finally, Aspinwall on the Panama isthmus. This was a traditional mission. At each stop, Rear Admiral Cooper reported carefully to the department economic information such as the port's main imports and exports and the main crops grown. He noted the number of ships in each port and how many of them were American. At each stop he was wined and dined by the local dignitaries, whom he invited in turn to be entertained on board *Tennessee*. Such socialization served to promote U.S. business interests, assuring local leaders and expatriot businessmen alike of the stability and security following the Stars and Stripes. The detail that Cooper went into in his official reports concerning commercial opportunities suggests that he considered such business dealings to be an integral part of his job as commander in chief.[57]

Tennessee arrived back in New Orleans on 13 March 1883.[58] Cooper had ordered the squadron to assemble there after their individual cruises. After two weeks in New Orleans, *Tennessee* departed in company with *Vandalia, Kearsarge,* and *Yantic*. The squadron carried out brief fleet tactical drills on their way from the mouth of the Mississippi River to Tortugas Islands, off of Key West, Florida. Again fires were banked and sails set after the squadron stood out to sea.[59] The four ships then proceeded north independently and rendezvoused at Hampton Roads in May. After replenishing stores and making minor repairs, Cooper intended to conduct a week or ten days of exercises, but the department ordered him back to New York. *Tennessee, Vandalia, Kearsarge,* and *Yantic* were to take part in the celebration of the opening of the Brooklyn Bridge on 24 May, with Rear Admiral Cooper as the senior Navy representative.[60]

The request to have the squadron present for the Brooklyn Bridge festivities was an example of the sort of Navy Department tasking that prevented the North Atlantic Squadron from conducting any tactical exercises for the remainder of 1883. Not only were there several public relations events to be attended, but unrest in Haiti called ships away as well. "At the request of the State Department," read Cooper's orders to *Vandalia*'s commanding officer, Captain Wallace, "the U.S.S. Vandalia under your command has been detailed to proceed to Port-au-Prince . . . to care for the interests of Americans during the present troubles in that island." Wallace's orders went on to require him to "afford such protection and security to the Americans residing in the Island as they may require of you."[61] Later in the year, *Vandalia* would be relieved by *Swatara*.[62]

The remainder of the squadron stayed busy with a variety of tasks. *Alliance* was ordered to the north to visit the various fishing ports and show the U.S. flag in the always-contested Canadian fishing grounds.[63] She would later be joined by *Vandalia* and *Swatara*.[64] In July Rear Admiral Cooper was ordered to take the flagship to La Guayra, Venezuela, to represent the United States at the unveiling of a statue of George Washington in Caracas.[65] Other engagements included the Newburgh, New York, centennial celebration in October and the celebration of the evacuation of New York by the British, held on 26 November 1883.[66] The majority of the squadron was able to reunite at that time, with *Tennessee,* as well as *Colorado, Saratoga, Jamestown* and *Yantic,* in attendance. Soon it was cruising season. In December Rear Admiral Cooper gathered the squadron in Hampton Roads and issued his orders for the various warships' West Indies deployments.[67] Although they spent some time together at various events, there is no evidence that North Atlantic Squadron units carried out any tactical exercises in the summer or fall of 1883. The dual nature of the functions expected of the squadron was evident. While there was initiative within the Navy Department, certainly from the Bureau of Navigation, to concentrate the squadron's warships and exercise them frequently, there was no set plan for executing this. Although there was a general idea that the squadron should cruise to the south in the winter and concentrate in the north during the summer, this convention was easy to ignore if exigencies arose. Senior officers also believed that they had, at that point, simply gotten everything they could out of having the wooden cruising vessels practice formations at the extremely slow speeds of 4–6 knots. As Admiral Cooper put it while at Hampton Roads in May: "I do not think it will be of any advantage to devote a longer period to these exercises as all the vessels have already had much practice in them."[68] Cooper's actions and his after-action reports clearly demonstrate that he felt that the North Atlantic Squadron had reached the outer limits of what productive good could be accomplished with wooden cruising ships.

Rear Admiral Luce and the Naval War College

One officer who was determined to push for a more systematic approach to fleet training and readiness for combat was then-Commo. Stephen B. Luce. Luce was the epitome of that rare breed of officer who was both exceptionally successful at sea and a pathbreaking leader ashore.[69] He, perhaps more than any naval officer of the nineteenth century, understood that a "fleet" was more than just a collection of ships. He both articulated and then put into action a comprehensive system of education. In 1841, as a fourteen-year-old, he signed on board the USS *Congress* as a midshipman and moved through the ranks over the next twenty years. After distinguished service during the Civil War, Luce's association with the North Atlantic Squadron began with his tour of duty as the commanding officer of Rear Admiral LeRoy's flagship, *Hartford,* from 1 November 1875 to 21 August 1877. Although he was not present during the Key West exercises of 1874, we know that he understood the importance of that initial set of maneuvers, since his personal papers contain a full set of copies of all the reports submitted by Rear Admiral Case.[70] Luce would, however, have been present for the landing exercises held in 1876.[71] As the commanding officer of the flagship, he would have been privy to Rear Admiral LeRoy's frustration that year as planned fleet tactical exercises off Port Royal were rendered impossible to carry out by Navy Department tasking, which scattered his warships throughout the North Atlantic Squadron's operating area. It was a pattern that would repeat itself during Luce's career with the North Atlantic Squadron: high hopes of executing fleet training undermined by other duties. After leaving command of *Hartford,* Luce turned to naval education and training. He successfully established the New York State Maritime School, then spent the years 1877–83 in various positions associated with training naval apprentices, including command of the U.S. Training Ship *Minnesota* and command of the Apprentice Training Squadron.

His interests extended to education for officers as well, which led to his most lasting contribution as the founder and first president of the Naval War College at Newport, Rhode Island. In *Professors of War*, Ronald Spector argues that the foundation of the NWC was an important step in the professionalization of the naval officer corps.[72] The opportunity for postgraduate professional interaction, when added to the initial bonding experience at the Naval Academy, the work of the Naval Institute at Annapolis, and the networking influence of various military-themed periodicals, was a major move for the profession. As such, the foundation of the NWC is a subject that has received its share of attention from naval historians. Typically the narrative focuses on the study of strategy and the cast of characters usually features Alfred Thayer Mahan and his ideas about the political-economic role of a navy in the shaping of national destiny. While correct, this interpretation does not give enough attention to Luce's belief in the importance of development of operational naval tactics in his fight to establish the Naval War College. Luce had a passion for putting naval theory into practice. He was fundamentally interested in the daily work associated with operating large ships together. One of the reasons that Luce felt that something like a war college was necessary was that the new naval professional of the 1880s would have to learn to fight entire squadrons of ships together as a unit.[73]

After much lobbying, on 3 May 1884, Luce was ordered, along with Cdr. William T. Sampson and Cdr. C. F. Goodrich, to "consider and report upon the whole subject of a post graduate school or school of application, to be established by the Navy Department for officers of the Navy."[74] The report that these three officers submitted the following year specifically noted, under the heading "PRACTICAL EXERCISES," "The North Atlantic Squadron affords the nearest approach to be found to a proper course in naval tactics. It should be assembled once a year, and during a stated period, go through a series of fleet evolutions, gunnery practice with the latest types of ordnance, the landing of seamen for military operations, boat operations, torpedo attack and defense, etc, having the class on board for instruction."[75] It is evident that from the beginning, the Naval War College was not intended by Luce to be simply a classroom-based institution. Before it had even been officially chartered, the NWC concept included the study and development of formation steaming tactics, with the North Atlantic Squadron acting as the laboratory.

On 26 July 1884, Commodore Luce was ordered to take command of the North Atlantic Squadron.[76] It was to be a temporary position, as Luce had already been tapped to open the new Naval War College later that year, but he was determined to make the most of his brief opportunity to command ships at sea. On 10 July 1884, *Tennessee, Vandalia, Alliance,* and *Yantic* got under way from the squadron anchorage off Staten Island and headed for Portsmouth, New Hampshire, where the change of command was to take place. Rear Admiral Cooper took advantage of having four ships steaming together to carry out one final set of fleet exercises under his flag. Moving out of New York Harbor in column, the ships commenced exercises at 9:10 a.m. Nine more days of exercises followed, until the squadron arrived at Portsmouth, New Hampshire, where they met *Swatara*. After conducting the change of command, the five ships of the North Atlantic Squadron, together with the ships of the Training Squadron, participated in the reception for the Greely Relief Expedition.[77]

From Portsmouth, *Tennessee, Vandalia, Swatara, Yantic,* and *Alliance* got under way on 6 August and conducted ten days of tactical exercises, including a landing of the naval brigade on Gardiner's Island on 11–13 August. From the ships of the squadron, 660 men were landed under the command of *Tennessee's* commanding officer, Capt. J. N. Miller. Luce proudly noted that it had been a surprise exercise, with the landing orders given after the squadron had left Portsmouth for Newport, and that it was the largest exercise of its kind ever conducted on as little notice.[78] The squadron arrived in Newport on 16 August.[79] Once in Narragansett Bay, Luce had the ships of his squadron conduct measured mile speed and tactical diameter tests.[80] Knowing exactly how many revolutions needed to be ordered for each ship to make a given speed, as well as knowing the arc each ship would scribe through the water at a given rudder angle, was crucial to the ability of a squadron to operate together and was information that was typically lacking at this formative stage. Officers of the deck were previously expected to carry out tactical maneuvers by "seaman's eye" rather than rely on data.[81] In the days before instrumentation, maneuvers were made much more difficult without a reliable way to know how fast each ship was going. By immediately ordering speed trials for his new command, Luce showed that he recognized this fact and he intended his squadron to spend a lot of time operating together. Even before the official opening of the Naval War College, Luce was doing his best to fulfill his vision of the North Atlantic Squadron as a squadron of evolution, working out tactical problems studied at the NWC through actual exercises at sea. In fact, the board's selection of Newport, Rhode Island, as the permanent location for the college had a lot to do with the fact that the proximity of the deep water of the Narragansett Bay made it easy for the entire North Atlantic Squadron to call at Newport and coordinate fleet exercises with the college. However, Luce's first tour as commander in chief of the North Atlantic Squadron was short lived, as Congress approved the secretary of the navy's recommendation to open the Naval War College based on the report of Luce's board. Naturally, he was tapped to be the first president of the college, which cut short—for the moment—his squadron command tour. In fact, much of Luce's correspondence during this period was focused more on his work to get the college up and operating than it was on his position as a squadron commander in chief.[82] On 20 September 1884, on board *Tennessee* anchored in Newport Harbor, Luce turned over command of the North Atlantic Squadron to Rear Adm. James E. Jouett.[83] He then went ashore to take possession of the abandoned poorhouse on Coasters Harbor Island and begin the work of establishing the Naval War College.

Rear Admiral Jouett and Intervention in Panama, 1885

James E. "Fighting Jim" Jouett entered the Navy as a midshipman in 1841, graduating from the Naval Academy six years later, in 1847. The highlight of his distinguished service during the Civil War was fighting alongside Rear Admiral Farragut at the Battle of Mobile Bay, as commanding officer of the Union steamship *Metacoma*.[84] Unlike the previous two changes of command of the North Atlantic Squadron, there was no grand review or series of tactical exercises. Jouett returned to New York with his new flagship *Tennessee,* where he remained until after New Year's Day. As 1885 dawned, the most important item on the squadron's calendar was representing the Navy at the World's Industrial and Cotton Centennial Exposition in New Orleans during the festive Mardi Gras season. On 10 January 1885, *Tennessee* arrived at Fort Monroe, Virginia, en route to New Orleans. However, on 4 March, Jouett received word that a revolution in Colombia threatened the transit of people and goods across the Panamanian isthmus. Contingency took precedence over squadron training opportunities from that point on. Jouett would be forced to spend most of 1885 on what would become the most noteworthy event of his career as a flag officer—responding to the crisis in Panama. *Galena,* with Cdr. T. F. Kane in command, was immediately ordered to proceed to Aspinwall—the Atlantic terminus of the Panamanian isthmus—with "all possible dispatch."[85] When Kane arrived on 13 March, he found an insurrection under way and the city of Aspinwall in danger. After communicating by telegraph with the Navy Department on 14 March, *Galena* was ordered to stay until further notice. Jouett did not leave New Orleans at first, but carried on with his entertaining duties. The situation did not appear to be overly serious, and his actions were in keeping with the usual role of a squadron commander in chief as a manager of scattered assets. Jouett even had time to request and receive permission to leave the flagship and travel to Washington, D.C., on personal business. The trip was no doubt timed to coincide with the assumption of office of a new secretary of the navy. On 4 March 1885, Democrat Grover Cleveland was sworn in as president. His new navy secretary was New York financier and political reformer William C. Whitney.[86]

The situation in Panama, however, was getting out of hand. Rebels set fire to Aspinwall on 30–31 March, destroying and damaging millions of dollars' worth of U.S. property. When the insurgents began burning the city, Commander Kane allowed American citizens to seek refuge in *Galena* while he sent his naval battalion ashore to protect U.S. property. When insurrectionists captured a mail steamer belonging to the Pacific Mail Line, Kane immediately recovered her.[87] He also arrested two of the more prominent insurrectionists and held them on board *Galena*. He let it be known that he was unwilling to turn them over to Colombian authorities, as their corruption or incompetence would allow the criminals to escape.[88]

Jouett was ordered to coal and proceed to Pensacola with *Tennessee*. There a detachment of sixty Marines boarded the flagship, which departed on 4 April headed directly to Aspinwall.[89] *Alliance* was sent to the vicinity of Cartagena, where her commanding officer, Cdr. Lewis Clark, was to make contact with U.S. consular officials and protect U.S. property and business interests.[90] *Swatara* was sent directly to Aspinwall to reinforce *Galena* as quickly as possible.[91] When Jouett arrived he took personal charge of the situation. His initial letter to the Colombian government representative at Aspinwall demonstrated the fine balance of military muscle and diplomatic tact that was required of a nineteenth-century flag officer. After announcing his arrival ("with *four* vessels of the United States Squadron under my command"), he went on to assure the authorities that he had no intention of interfering with the constitutionally recognized government of Colombia and requested permission to land additional U.S. troops if he deemed it necessary. It is doubtful that anyone reading the letter, least of all the Colombians, was under any illusion that Jouett cared about their permission, but the diplomatic niceties were observed to the letter.[92] Under Jouett's supervision, transit across the isthmus reopened on 11 April. By the fourteenth Jouett was able to report to Secretary Whitney that the situation had been stabilized.[93]

On 15 April a naval brigade consisting of sailors and Marines under the command of Cdr. B. H. McCalla arrived from New York to assist Rear Admiral Jouett in restoring order and "maintaining treaty obligations" (i.e., keeping transit across the isthmus open for U.S. commerce).[94] Jouett gave McCalla careful instructions to as much as possible not interfere with internal Colombian politics. He then sent McCalla's force on the railroad across the isthmus to prevent Panama City from being burned as Aspinwall had been. In light of the orders he had given, Jouett was taken aback when McCalla formally occupied the entire city. He quickly instructed McCalla to remove his troops to the train station and worked to smooth relations with Colombian officials.[95] In any event, Panama City was spared the fate Aspinwall had suffered two weeks earlier. By 24 April, Jouett was able to report that all was quiet, and that a contingent of seven hundred Colombian troops was expected to arrive soon, in which case Jouett planned to withdraw McCalla's troops and turn Aspinwall and Panama City over to the proper Colombian authorities.[96] Once the Colombian troops arrived, Jouett, with two officers of his staff, rode the railroad across the isthmus to Panama City to meet with them and personally express his support for the constitutional government. Jouett was well received by the Colombian officers in charge of the detachment, who were grateful for his assistance and assurances about Colombian sovereignty.[97] In that spirit, they asked Jouett to deliver the two prisoners Commander Kane had taken on board *Galena* after the burning of Aspinwall.[98] Jouett assented to the prisoner transfer. On 7 May, after a quick court-martial, the two were publicly hanged in Aspinwall. Jouett pronounced the outcome "beneficial."[99] On 8 May the first contingent of U.S. Marines boarded a transport for home, as more Colombian troops arrived to secure the city.[100]

With the military situation secure, at least in regard to U.S. interests in the isthmus, Rear Admiral Jouett turned his attention to diplomacy. On 11 May he proceeded in *Tennessee* to Cartagena, the capital of Colombia, for the purpose of mediating a permanent cession of hostilities between the rebel forces and the forces of the constitutional government.[101] There *Tennessee* happened upon two steamers loaded with insurrectionists who were planning to retreat after having been repulsed during a battle for the capital city. Jouett refused to let the rebels leave and informed them that he would prevent the departure of their vessels from Cartagena's harbor by force if necessary. He then invited the rebel leaders to join him in *Tennessee,* where he persuaded them to allow him to attempt to mediate a settlement. Unfortunately, by 25 June, Jouett was writing the Navy Department that "a peaceable settlement" would be impossible.[102]

Although they did not have an opportunity to conduct fleet maneuvers, at one time or another the entire North Atlantic Squadron was involved in the Panama operation. *Swatara* remained in Colombian waters until July, at which time she proceeded back to the United States. *Alliance* remained for two months, departing in June. *Yantic* arrived in May, sailing from Guatemala, where she had been ordered to protect U.S. interests at Livingston. She relieved the other ships and stayed until 1 August.[103] Rear Admiral Jouett and *Tennessee* remained until 11 July, when with a yellow fever outbreak threatening the health of his sailors he was ordered north by the Navy Department.[104] *Tennessee* arrived at Fort Monroe, Virginia, on 23 July 1885.[105]

The Navy Department was sensitive to public opinion surrounding the Panamanian operation. Democrat Grover Cleveland became president in March 1885, replacing Republican Chester Arthur. Cleveland had run on a platform of non-intervention and disapproval of the aggressive foreign policy and expansionist tendencies of Arthur and his secretary of state, Fredrick Freylinghuysen. It would not do to have the first foreign crisis to confront his administration be a naval intervention that got out of hand. Before his departure with the troop reinforcements, Commodore Walker, the chief of the Bureau of Navigation, reminded Commander McCalla that "it is of considerable importance . . . that we keep the country with us in this matter." He went on to instruct McCalla to be sure to take every opportunity to send information back to the department, "that it may be given out to the press, and the people kept in accord with the Department."[106] This was apparently news to Rear Admiral Jouett, who when he opened the stateside newspapers that arrived in Aspinwall on 1 June was incensed to see that correspondence between Walker and McCalla to which he was not privy had been published for the general public. Furthermore, some of McCalla's information and opinions directly contradicted information given by Jouett in his own official dispatches. Jouett demanded, and received, from McCalla copies of every communication he had had with the Navy Department and requested clarification of McCalla's subordinate role from the secretary of the navy.[107]

In the meantime, a letter arrived from the commanding officer of the USS *Wachusett,* who was none other than Alfred Thayer Mahan. Mahan, whose world fame was years in the future, had been sent to Panama City by the commander in chief of the Pacific Squadron. His predecessor on station, Captain Norton of the *Shenandoah,* had carried out some tasks "suggested" by Rear Admiral Jouett in support of his mission to keep the isthmus transit open. Mahan now asked Jouett directly if he had the authority to order these tasks or if they were simply advisory in nature, in which case he did not intend to carry them out. Jouett wrote Mahan a curt reply ("I do not care to discuss the matter with you") and referred the whole matter to the secretary of the navy.[108]

Taken together, these two incidents demonstrate the structural difficulty of determining the operational chain of command in the era before the establishment of the Office of the Chief of Naval Operations. With each bureau its own entity, answering only to the secretary of the navy, Commodore Walker had no problem corresponding directly with Commander McCalla, even though it violated the chain of command at the scene in Panama. For his part, Mahan's letter was technically correct but short on political savvy. The secretary of the navy later agreed that the senior officer present, regardless of squadron, should be in charge of all matters at the isthmus. Mahan's predecessor on station had been much more politically astute, even if not as by the book, which shows why Mahan was considered by his contemporaries to be, at best, a mediocre line officer. He would go on to have a much more successful career as an academic.[109]

The operations of the North Atlantic Squadron in the spring and summer of 1885 show the simple effectiveness of the old way of dealing with traditional U.S. foreign relations problems—threats to property and the transit of goods and services. These threats were adequately addressed with the presence of one or more wooden ships in the port of Aspinwall and sailors and Marines deployed ashore. It would seem that the opinions of the majority report of the first Naval Advisory Board were justified, as Jouett's wooden cruisers carried out their mission in a timely manner with great effectiveness. Naval officers such as Lieutenant Belknap feared, however, that these capabilities would not be enough for a future encounter with a peer naval competitor. This seemed more likely as U.S. assertion of claims to exclusive leadership in the Western Hemisphere grew stronger throughout the 1880s.[110] The North Atlantic Squadron had to be able to keep the Panamanian isthmus open for business not only in the face of poorly armed insurrectionists but also in the event of hostilities with a European naval power. It did not help matters that a French national (Ferdinand de Lesseps) was then engaged in an attempt to build a canal across the isthmus. To meet future threats, either from South American nations or from European incursions into the Western Hemisphere, the North Atlantic Squadron would have to be capable of engaging an armored fleet at sea as a tactical combat unit.

The arrival of the flagship *Tennessee* back at Fort Monroe on 23 July 1885 coincided to the day with the death of former president Gen. Ulysses S. Grant.[111] Jouett had been trying to arrange some liberty for the crew of his flagship and the department had been anxious to have the squadron conduct some fleet tactical drills, but all was put on hold pending Grant's funeral.[112] Naturally, it was expected that warships from the North Atlantic Squadron would be present at the ceremonies in New York, which they were.[113]

The remainder of the year proved frustrating for Rear Admiral Jouett. Plans for squadron tactical exercises, desired by both Jouett and the Navy Department, were consistently hampered by the poor material condition of the squadron's wooden ships. After working through various mechanical conditions, Jouett had managed to collect three of his ships, *Tennessee, Alliance,* and *Galena,* at Bar Harbor, Maine, in August. Just as they were about to get under way, however, the Navy Department ordered *Tennessee* back to New York to have her seams recaulked.[114] Later in the fall, a series of exercises Jouett had planned to hold in Florida Bay were placed on indefinite hold by the Navy Department, and *Tennessee* was instead sent, along with *Galena,* on a cruise in the West Indies.[115]

Having thus been twice denied the opportunity to conduct fleet exercises in 1885, Jouett was determined to do better in 1886. He ordered his forces to rendezvous in Key West at the end of March 1886, for a week of fleet tactical exercises.[116] These were carried out between Key West and Pensacola during the month of April. While in the south, Jouett corresponded with Commodore Walker of the Bureau of Navigation about ways to keep his force intact. Walker's answer was instructive: "If you come north in the usual way, your ships are sure to be scattered to the different yards, and you will lose control of them just as you did last summer."[117] This was exactly what happened. Against Walker's advice, Jouett took the squadron, consisting of *Tennessee, Brooklyn, Swatara, Galena,* and *Yantic,* to New York, where they were promptly split up. *Brooklyn* went into the navy yard for work; the rest were sent to visit ports in the Northeast and Canada.

Rear Admiral Luce and the North Atlantic Squadron

Meanwhile, between October 1884 and June 1886, Stephen B. Luce had been busy at work at the Naval War College, honing his ideas about tactical theory and operational practice. Along the way he gathered associates, such as Admiral of the Navy David Dixon Porter and Commo. John G. Walker, who agreed with and supported his vision of postgraduate centralized education for naval officers.[118] He also provoked opposition, which included the superintendent of the Naval Academy, who viewed the NWC as an infringement on the academy's mandate as the home of officer education.[119] Superintendent Ramsey failed to understand the operational aspects of Luce's project. At the NWC, it was always Luce's intention to marry intellectual efforts at the shore establishment with practical work at sea.[120] He, perhaps more than any other officer of the time, understood that the complicated multiship formations that would characterize naval warfare of the future would require a different kind of naval officer. It would not be enough for these officers of the future to have a common entry-level education at Annapolis. They would require a new, more specialized body of professional knowledge, and this body of knowledge would have to be the same across the fleet, because these officers would be required to operate their ships in close formation and fight as multiship units. In short, modern naval warfare required the U.S. Navy to develop fleet doctrine.

To operationalize this, Luce brought to the NWC faculty retired lieutenant William McCarty Little, a member of the Annapolis class of 1866. While Alfred Thayer Mahan is the most famous of Luce's appointments to the college, McCarty Little would have to be a close second.[121] He had become acquainted with Luce while serving as the navigator on board the USS *New Hampshire*, one of the vessels in Luce's Training Squadron. McCarty Little's promising career had been cut short by a chronic eye condition that had cost him the sight of one eye and periodically threatened the sight in the other. Profoundly disappointed over his medical retirement, he enthusiastically joined the college staff, often on a volunteer basis without pay. While Mahan and Luce got much of the press for their publications and their sweeping ideas about national maritime strategy, McCarty Little quietly went about developing the methods for college students to try out steam tactics. When it proved impractical to gather enough actual ships in Narragansett Bay to conduct exercises, it was McCarty Little who suggested that the college use steam launches instead.[122]

After seeing the Naval War College safely established, Luce was sent back to sea in June 1886, this time as the permanent commander in chief of the North Atlantic Squadron.[123] As has been previously shown, Luce felt that the squadron that operated regularly on the East Coast should regularly work in conjunction with the NWC to try out tactics that had been developed by the college's students. Leaving Capt. A. T. Mahan—by now detached from *Wachusett* and established in

Newport—in charge of things ashore, he had successfully maneuvered to place himself in a position to be the important practical partner of the NWC's theoretical effort. He set to work immediately, corresponding with the Bureau of Navigation about what warships he would have assigned to him in the summer of 1887 and what he would be able to do with them.[124] As one of the junior officers in his squadron later recalled, "We immediately ceased to spend the summers at the principal New England watering places and the winters at the New Orleans Mardi Gras, and went into the most intensive and, as many learned to think, irritating and unnecessary tactical maneuvers."[125] Luce had another innovation that rankled his officers but gave a clue to the direction that the professionalization of the officer corps was taking: He liked to score his subordinates on their proficiency in carrying out tactical maneuvers, and he ranked them accordingly with these scores.[126] Promotion in the late 1880s was not yet done on the basis of merit, but officers such as Luce understood that if ships were going to fight together, their officers would have to be held to a common standard across the squadron.

Despite the support of the Navy Department, tactical exercises soon took a back seat to international politics. In November 1886, Luce was directed to send a warship back to Aspinwall, once again protecting U.S. citizens, business interests, and free transit of the isthmus during the continued political unrest there.[127] The following year would bring more tasking from the State Department, this time on the other side of the squadron's area of responsibility. Much of Luce's attention that year was directed to the Canadian fisheries question. In July 1887 the secretary of the navy ordered that the North Atlantic Squadron proceed into the Gulf of St. Lawrence to enforce the fishing rights of American fishermen, in accordance with the 1818 treaty establishing those rights and the 1871 Treaty of Washington. Unrest had been brewing over what was seen as unlawful British prosecution of U.S. fishing captains in Canadian waters.[128] The flagship was ordered to proceed to Portland and Halifax, while the *Galina* and *Swatara* were sent one at a time into the gulf.[129] Luce's handling of the fisheries controversy shows his matter-of-fact approach to such political questions and his desire to concentrate on preparing for what he felt was the true calling of the navy. Although he does not directly say so in his correspondence, one gets the feeling that Luce viewed these deployments of his ships as a distraction at best and an outright misuse of resources at worst. Rather than honing tactics to be used in naval warfare on the high seas, he was forced to spend much of the prime exercise season looking after the business interests of American fishermen. Contrasting Luce's correspondence with that of Rear Admiral Cooper (CinC, 1882–1884), whose dispatches were always newsy and full of commercial information, throws the two distinct, and often conflicting, missions required of commanders in chief during this era into stark relief. Although Luce had a vision of a complete system of training and exercise for his command, and although he had colleagues such as Comm. John G. Walker (about whom more in the next chapter) in positions of importance such as the chief of the Bureau of Navigation, the North Atlantic Squadron was still captive to the need to perform political missions as requested by the Department of State.

The North Atlantic Squadron returned to Narragansett Bay, where in conjunction with the U.S. Army it carried out a very successful series of landings and maneuvers in the fall of 1887.[130] Back at New York for the winter, Luce worked on fleet training plans for the following summer. This is a significant indicator of progress in the development of a multiship fighting capability. In contrast to ad hoc deployments of single ships based on requirements to "show the flag" and protect commercial interests, Luce's actions during the winter of 1887–88 show a commander in chief actively planning combat training for his squadron and working to incorporate that training into his unit's deployment plans. Even the previously unprecedented tactical training under Rear Adm. George Cooper in 1882 does not really appear to have been more than taking advantage of the squadron's orders to be present together at the various celebrations they participated in that summer.

From New York, Luce corresponded with General Sheridan of the Army, suggesting that the Marines participate in joint Army-Navy exercises in 1888.[131] He also carried on a regular dialogue with Commodore Walker and the secretary of the navy about available ships and their possible ports of call for the next summer.[132] The initial plan was for Luce to take his warships, in company, on a tour of the southern ports, namely, New Orleans, Mobile, Pensacola, Savannah, and Charleston, then proceed north. It made sense, the squadron having spent the last two summers visiting northern ports. It appeared that Luce would have *Richmond* (his flagship), *Atlanta* (the first of the New Steel Navy cruisers), *Yantic, Dolphin,* and *Galena.* This was not a large squadron, but there were enough ships to work through some tactical problems and train the officers of the squadron in handling their ships in formation.

The first indication that Luce was not going to be able to carry out his planned exercises in the summer was a request for support from the U.S. minister at San Domingo.[133] On 11 January 1888, only ten days after Walker had expressed the approval of the Navy Department for Luce's training plan, Walker wrote Luce a somewhat apologetic letter in which he instructed him to detach a ship to serve the needs of the State Department.[134] In July Luce's flagship *Richmond* was summoned for service on the Asiatic station. He was given *Pensacola* as a replacement, but she was unseaworthy, so he would be forced to transfer his flag to another, smaller, ship if he wanted to lead at sea.[135] Meanwhile, political conditions in Haiti[136] were deteriorating throughout the summer, culminating in an order from the Navy Department in August to send a ship to Port-au-Prince.

About this time, a letter arrived from the Navy Department asking Luce's opinion on summer training plans for his squadron. It should have been obvious at this point to anyone bothering to pay attention in the Navy Department that Luce had at his command only two ships. Training of any sort, other than perhaps to send signals to one another, was completely out of the question. The letter was the last straw. On 28 July 1888, Luce fired off a seven-page reply from New York in which he described his attempt to put together a coherent training plan for that summer. He detailed the detachment of his ships, one by one, for tasking to support the State Department, and he questioned, with astonishment, the attempt by the Navy Department to charge the War Department for any coal expended carrying soldiers in Navy ships during combined exercises. He lamented his inability to carry out his vision of making the North Atlantic Squadron a "school of practical instruction" that would exercise the theoretical concepts developed by the Naval War College:[137]

> The *fundamental idea* [emphasis added] is to make theoretical instruction and practical exercise go hand in hand; or, in other words, to correlate the work of the Squadron and that of the College. In the lecture room certain tactical propositions are laid down, or war problems given out, to the officers under instruction. Their merit is then tested in the School of Application, the Squadron, and the result afterwards discussed in the lecture room. This system raises our Squadron exercises to a higher plane than those of any other known to me, and places our Navy, comparatively insignificant in all else, in advance of the Navies of the world in respect to professional education.[138]

The 28 July 1888 letter from Luce to Secretary of the Navy Whitney is a pivotal piece of Luce's correspondence, second only perhaps to the letter inviting Alfred Thayer Mahan to join the faculty of the Naval War College. Here, encapsulated in one document, is the basic difference between the modern fleet concept and the historical utilization of the U.S. Navy. Under Luce, the identity of the North Atlantic Squadron was that of a single combat unit, which sailed together, trained together, and expected to fight together. In short, the North Atlantic Squadron was an embryonic fleet, in the modern use of that word. To the State Department, however, and to a lesser extent the Navy Department, the North Atlantic Squadron was simply a collection of ships, from which the executive branch could draw upon as necessary to fulfill commitments to U.S. citizens, property, and business interests throughout their area of responsibility. While squadron exercises became commonplace, and even expected, throughout the decade of the 1880s, it was clear in 1888 that the new concepts had not yet been accepted as the basis of peacetime naval operations.

In any event, Luce did not have long to stew about his failure to convince the Navy Department of the validity of his views. Down in Haiti, the political unrest that had already deprived him of one of his ships earlier in the year had taken a turn for the worse. The *Haytian Republic,* a steamer flying the U.S. flag, was seized by the Haitian government. This was a clear violation of the international rights of U.S. citizens, and one that struck especially at the sensibilities of a United States always keenly interested in the protection of U.S. property abroad. On 8 December 1888, Luce was given back *Richmond* (temporarily) and told to take her and his remaining two ships *Galena* and *Yantic* and depart for Port-au-Prince at once. *Ossipee* would meet them on the way down, as they passed Norfolk. In the event, Luce accomplished the job with only *Galena* and *Yantic,* the other two vessels not being ready for sea fast enough. There is little doubt that Luce thought that there was a good possibility that hostilities would result, as he drilled his little command and made out battle instructions while in transit.[139] The two ships would prove to be enough, however. They entered the harbor at Port-au-Prince at quarters, cleared for action with guns loaded. The provisional government, sensing that this was a fight that would be unprofitable for them, quickly released the *Haytian Republic.* In a letter to Secretary of the Navy Whitney, Secretary of State Bayard praised the "high and intelligent discretion which has characterized the action of Admiral Luce in the execution of this National duty to American citizens."[140]

Conclusions

Luce applied for and received his detachment from the North Atlantic Squadron in January 1889. The decade of the 1880s had seen a change in the North Atlantic Squadron, not in structure or matériel, but in its sense of itself as an organization. As in any organization undergoing a fundamental change in image and identity, the squadron inhabited a middle ground between the old identity and the new. Although this characterization of the squadron's dual identity would be accurate until the middle of the 1890s, it was at no time truer than during the 1880s. The command tours of Rear Admiral Cooper and Rear Admiral Jouett bring this characterization into relief. Under Admiral Cooper's somewhat reluctant leadership, the squadron carried out at least four major sets of exercises, operating as a unit for a total of forty-four days. But the tactical exercises under Cooper were not part of an overall plan readying the squadron for combat as a tactical unit. They were products of opportunity that were dropped as events occurred that were determined to be more important to the squadron's critical function of showing the flag and protecting and promoting U.S. commerce. This is seen clearly in the command tour of Rear Admiral Jouett, who was only able to conduct a single week of tactical exercises, in April 1886. The highlight of his tenure as commander in chief was the revolution in Colombia, an experience very much in keeping with the old Navy image of the naval officer as a warrior-diplomat.

After Rear Admiral Luce took command of the squadron in 1886, he brought a vision for an integrated training plan. Under his leadership, the North Atlantic Squadron warships not only trained together more often but did so as part of an overall scheme linking the theoretical work of the Naval War College with practical preparation. A routine developed that sent the warships of the squadron north in the summer so that their officers could participate in the college's summer session then return to their ships to put into action theoretical concepts worked out in the classroom. After these summer exercises, the squadron could send warships north to the Canadian fishing waters or south to the Caribbean. Its identity was becoming more that of a fighting unit and training organization and less that of an administrative body that facilitated assignment of ships to individual missions by the Navy Department. Years before Mahan popularized the theory of seapower, the operational patterns of the North Atlantic Squadron were laying the foundations for the development of a national battle fleet.[141]

This vision was only partially realized in the 1880s.[142] The decade to come would bring not only matériel changes with the arrival the first of the steel ships of the New Steel Navy but also changes in the way those ships were employed. It would also bring to the forefront a powerful, politically connected officer who shared Luce's vision for a well-trained fighting squadron. Under Rear Adm. John Grimes Walker, the Squadron of Evolution, consisting of the *Atlantic, Boston, Chicago,* and *Dolphin,* would tour Europe, showing the nations of the Old World that the naval power of the United States was in the process of rebirth.

3

The North Atlantic Squadron and the
Squadron of Evolution, 1889–1891

This chapter traces the trailblazing deployment of the newly established Squadron of Evolution, from November 1889 to the fall of 1891, in the context of the day-to-day operations of the North Atlantic Squadron. Comparing the leadership styles as well as the operational employment of the two squadrons gives examples of the slow development of the organizational identity of the North Atlantic Squadron as the nation's foreign policy became more coherent and the U.S. Navy shifted from a focus on cruising and commerce raiding to one of engaging enemy fleets in open-ocean combat. The Squadron of Evolution will be shown to be the operational expression of the new identity that had been slowly coalescing in the North Atlantic Squadron over the previous decade.

The 1888 presidential election brought the Republican Benjamin Harrison into office. This paved the way for James G. Blaine to return to the position of secretary of state, a position he had held a decade earlier under the Garfield administration. Blaine's emphasis on U.S. trade initiatives and presence in Latin America meant increased commercial activity in the North Atlantic Squadron's chief area of operations. These commercial interests spurred a renewed concern about a canal across either Nicaragua or the isthmus of Panama. It also meant that the United States became increasingly concerned about commercial forays into Latin America by European governments and the possible implications that could have for the Monroe Doctrine. The activist, sometimes outright belligerent foreign policy of the Harrison administration, coupled with a favorable economy and public interest in expansionism, is perhaps best captured in historian Frederick Jackson Turner's famous frontier thesis, presented in 1893.[1] All of these foreign policy developments meant that the North Atlantic Squadron would stay busy in the decade of the 1890s. Busyness, however, did not necessarily translate into combat readiness. This highlighted the never-ending conflict between the desire of the State Department to have U.S. warships protect commercial interests and the ability of the squadron to exercise as a combat unit.

Rear Admiral Gherardi and the North Atlantic Squadron, 1889

With the successful outcome of the *Haytian Republic* affair barely behind him, Rear Adm. Stephen B. Luce was detached from command of the North Atlantic Squadron on 28 January 1889.[2] He was relieved at Key West by Rear Adm. Bancroft Gherardi, whose previous assignment had been as commandant of the New York Navy Yard.[3] It is significant that in commenting on this, both the *New York Times* and the secretary of the navy report for 1889 noted that "the *squadron* on this [North Atlantic] station is now under the command of Rear Admiral Bancroft Gherardi."[4] The image of the North Atlantic Squadron was slowly transforming, as outsiders (in this case, newspapers as well as the secretary of the navy) began to use language consistent with viewing the squadron as a fighting unit. This can be contrasted to reports from as late as four years earlier that referred to the "force on this station."[5]

The commander in chief's flag was still flying from *Galena* when Rear Admiral Gherardi took charge. *Galena* remained at Key West for another two months, getting under way again in February for the Caribbean. Unrest in Haiti continued to occupy the attention of the North Atlantic Squadron throughout much of 1889. At issue was the longstanding desire of the United States to have a naval presence on the island of Hispaniola. The dueling forces of François Legitime (president from October 1888 to August 1889) and Florvil Hyppolite (president from October 1889 to March 1896) were attempting to disrupt the flow of arms and supplies to each other's supporters. On his previous visit to the area in 1888, Rear Admiral Luce had determined that President Legitime's gunboat navy did not have the resources necessary to establish a legal blockade. Their declaration of a blockade was therefore illegitimate. *Yantic* had then remained behind after the *Haytian Republic* affair had concluded and Luce departed. She maintained a presence on station until a case of yellow fever forced her return north in January 1889. The Haitian attempts at blockade were a disruption of business in an area considered to be strategically important and were thus a matter of continuing concern for the Navy.[6] After a cruise through the area, *Galena* returned to New York in May 1889.

In New York, Rear Admiral Gherardi was ordered to transfer his flag to the venerable *Kearsarge* and then return to the Caribbean for another cruise in Haitian waters. *Kearsarge* was added to the North Atlantic Squadron to replace *Ossipee,* which was scheduled for decommissioning at the end of the year. Gherardi had been offered *Boston* as his flagship, but he demurred, preferring the older but much more spacious *Kearsarge.* Much of the available interior room in the new steel warships was taken up by machinery; their living conditions were considered cramped, even without the added personnel of an admiral's staff. The older wooden cruising vessels, even a sloop like *Kearsarge,* offered plenty of room for a flag staff's operations in addition to the ship's company.[7] The relatively minor issue of the selection of a flagship seems a small point, but it is not. It provides evidence that Rear Admiral Gherardi did not view his command the same way that Luce had. The idea of Stephen B. Luce turning down the opportunity to fly his flag on one of the first ships of the New Steel Navy would have been unthinkable. Gherardi's selection of a obsolescent wood sloop as his flagship shows that he was more focused on having the room necessary for his staff to administer the squadron's presence duties than he was leading the squadron in multiship tactical drills.

It did not escape the attention of the *New York Times* that the North Atlantic Squadron had not exercised in fleet tactics since the departure of Rear Admiral Luce, the newspaper going so far as to note, "It is true that not a few commanding officers of vessels dislike squadron operations. This became eminently conspicuous during the rumored fitting out of a 'flying squadron' designed to cruise round the world. To have their vessels assigned to such a squadron would completely handicap the independence of the Captains, for henceforth their every movement would be regulated by the will of the Admiral in command of the fleet."[8] This is exactly what, for the moment, was precluded in the North Atlantic Squadron, but not in the soon-to-be-constituted Squadron of Evolution.

Kearsarge's next assignment caused some controversy and illustrated the identity changes taking place in the nation, as well as the Navy. In September she was detailed to carry the newly appointed U.S. minister to Haiti—Frederick Douglass. *Ossipee* had originally been given the assignment, but the Norfolk Navy Yard reported that her boilers were in need of two weeks' worth of repairs before she would be seaworthy, so *Kearsarge* was given the assignment instead. Newspapers in New York and Washington picked up the story and reported that *Ossipee's* captain and executive officer were uneasy about the social status of the African American dignitary they were ordered to carry on board and had fabricated the mechanical problems to avoid the duty. This was denied vehemently by everyone involved, including *Ossipee's* executive officer, Admiral Evans, in his memoirs.[9] Appearances were not helped when the commanding officer of the *Kearsarge,* Commander Shepard, was quickly relieved by Commander Whiting the next day. The official explanation was that Commander Shepard had previously asked to be relieved from sea duty, but the timing of the change of command pointed to the possibility of his sharing the same racial sensitivities allegedly attributed to the *Ossipee's* commanders.[10] Eventually Minister Douglass, his wife, daughter, and private secretary were housed as comfortably as possible in Commander Whiting's cabin onboard *Kearsarge* and delivered to Haiti without further incident, arriving on 8 October 1889.[11]

Admiral Walker and the Squadron of Evolution

Meanwhile, the first steel ships of the New Steel Navy were making their operational appearance. It had been a long time coming. After all the work done by the two naval advisory boards in 1882 and 1883, Congress eventually authorized the construction of three steel cruisers and a dispatch vessel. Hunt and Chandler provoked much debate in Congress. The fact that the Rodgers board had reported out both a majority and a minority report did not exactly achieve the unified vision for a national naval strategy that Secretary Hunt and later Secretary Chandler had hoped for. Nonetheless, on 3 March 1883, Congress appropriated money for the construction of three protected cruisers and a dispatch vessel. The contract for all four of the ABCD ships was awarded to John Roach's shipyard in Chester, Pennsylvania. Roach was a friend of Secretary of the Navy Chandler, a fact that caused no small amount of public furor, but, in fact, he had the only facilities capable of handling steel of the amount and size necessary to construct steel warships.

The Navy's first steel warship, *Atlanta,* was commissioned in 1886. She and her identical sister ship, *Boston,* were both 270 feet long, carried a crew of 265 enlisted personnel and 19 officers, and boasted two 8-inch rifles, six 6-inch rifles, and a battery of various smaller weapons. The flagship of the Squadron of Evolution, *Chicago,* at 325 feet, was the largest of the four and generally considered the best looking.[12] She drew 4,500 tons and had a crew of 376 enlisted and 33 officers. Her armament consisted of four 8-inch guns, eight 6-inch guns, two 5-inch guns, and various other quick-firing smaller weapons. The three cruisers joined the 1,500-ton dispatch vessel *Dolphin,* which had been the first of the ABCD ships launched.

These four ships represented halfway points between the wooden steamers of the 1870s and the modern ships that were to come in the next twenty years. While designed with double steel hulls, watertight compartments, and fully electric lighting systems, they retained masts, canvas, and the ability to make way under sail power with partial sail rigs. As "protected cruisers," these were essentially unarmored ships. They had a thin layer of steel plating that covered the top of the vital engineering spaces but no armor belt along the sides. With the exception of *Chicago,* they had single screws, underscoring the fact that they had not been designed for extensive formation work. The axial forces generated by the rotation of a single screw on the centerline require constant rudder corrections to prevent the vessel from constantly falling off course. This makes a single-screw ship significantly more difficult to maneuver precisely than one provided with two screws, one on either side of the keel. The construction of these four ships gives an insight into the expectations of Congress and the Department of the Navy when they approved the designs of these ships. Although modern in many respects, these were still cruising vessels, not designed for the line of battle. "A solitary American steel cruiser with its delusive prefix of 'protected,'" wrote Stephen B. Luce in 1889, "represents the latent possibilities of a great country placidly awaiting some national disaster to generate its mighty forces."[13]

By the spring of 1889, the first three cruisers of the New Steel Navy were almost ready for squadron assignment. How the new ships were to be assigned and utilized was the subject of much speculation, both among naval officers and the general public. The assumption was that the new ships would be spread out among the established stations around the world. In other words, exactly as the wooden steamships they were replacing were used. However, in May 1889, stories began to be whispered about other plans that the Navy Department might have. Soon it became evident that the three new steel cruisers were to be kept together as a squadron.[14] This decision was due in large part to a naval officer who would continue Stephen B. Luce's movement to change the way the ships of the navy were employed. As the chief of the Bureau of Navigation and Detail, Commo. John G. Walker enjoyed the confidence of the secretary of the navy. He also had an inordinate amount of power over the movement of ships and the detailing of officers to man them. After two consecutive tours in the bureau, Commodore Walker was in line to be detailed to sea. The command he desired was the North Atlantic Squadron, but that command had been promised to Rear Admiral Gherardi, who was senior to Walker. Unable to get himself placed in the "twilight tour" he desired, Walker went about quietly setting up the next best thing: command of the navy's newest vessels. Walker was known within the department as someone interested in concentration of naval assets whenever possible, and we have seen him encourage Jouett and Luce to keep their ships together and conduct exercises.[15] In July 1889, while actively lobbying for the formation of the "White Squadron," he persuaded the secretary of the navy to appoint him the head of a board that would conduct trials on the new cruisers *Atlanta, Boston,* and *Chicago.* The so-called Walker Board would be responsible for determining and documenting the maximum speed and horsepower generated, as well as the turning radiuses, of the three ships.[16] This information was of particular importance for two reasons. The first was to establish exactly what top speed could be expected from the ship operationally. The structure of the contracts to build the ships promised bonuses for the shipbuilder for excess horsepower developed and knots of top speed. In pursuit of these bonuses, the shipbuilders conducted acceptance trials with the ship crewed with the best stokers and firemen that money could buy, using the highest-grade coal. Naval officers had good reason to be suspicious of these results and wanted to run trials for themselves using ordinary sailors as the crew and standard-grade coal for the boilers. The second reason this information was important was that although the new steel ships had been designed as cruisers—optimized for single-ship operations—it was becoming evident that the Navy intended to conduct multiship squadron operations with them. In the absence of modern instrumentation, which gave accurate readings of ship speed and rudder angle, it would be necessary to take careful note of their speed and turning radius through experimentation.

Accordingly, as they were completed, the three cruisers made their way to the Narragansett Bay and the waters off of Newport, Rhode Island, for the trials.

Within the decentralized bureau system of the Department of the Navy, the unusual appointment of a board to collect this information caused annoyance if not offense. The Board of Inspection and Survey, headed by Rear Admiral Jouett, a previous North Atlantic Squadron CinC, was supposed to conduct all trials for new warships. Jouett, although on friendly terms with Walker, resented that well-connected officer's intrusion onto what he considered to be his turf. Jouett's entreaties to the admiral of the Navy were dismissed, however, and the new ships continued with their trials under the supervision of the Walker Board throughout the fall of 1889.[17]

Meanwhile, Walker was looking ahead to the work he planned to do while under way with his new squadron. In preparing for deployment, he had to confront a Navy structure that was unprepared to support the innovative work he was trying to accomplish. Methods of signaling provide an example of this. In the deployments analyzed in the previous chapter, signals exchanged between ships were few, and routine in nature. When ships were together in port, the senior officer present would coordinate the raising and lowering of topgallant yards and occasionally request that junior ships "send a boat" to receive instructions. None of this required dedicated signals personnel. The officer of the deck and whatever sailors on watch were assisting him could handle the duties of decoding these flags and reporting their meaning to their captain for action. Walker, however, had different things in mind, and he knew that constant and rapid communications between ships would be a vital requirement. To that end, he instructed each of his captains to select six especially capable sailors and train them to handle signal flags.[18]

Being prepared for daytime signaling was not enough in Walker's mind. In a letter to the secretary of the navy, written just prior to departing New York, he complained that the ships' allocation of rockets and Very signals was not adequate for the "amount of night signaling which I propose to do in this Squadron."[19] For all the formation work that previous commanders in chief had been successful in carrying out, only a small fraction had been done at night, mostly to evaluate new night signaling devices.[20] Walker, who contemplated his four ships spending the vast majority of their deployment in company, found that the Navy Department bureaucracy that supplied the Navy's warships had not caught up with the plans for a Squadron of Evolution.

By and large, the media approved of these plans. "It is now patent," crowed the *New York Times*, "that Admiral Luce's ideas were proper ones in the matter of handling squadrons, and were furthermore the only right ones for the securing of efficiency on the part of naval forces when called upon for duty ashore."[21] The old days of single ships under sail were giving way to concentrated multiship operations under steam. This required a different set of competencies, and naval officers—especially young ones—applied to the Bureau of Navigation and Detail for a chance to be a part of this cutting-edge experience.[22] It was as if naval officers of the era knew that the future had arrived, and that future was not duty on a single ship showing the flag by itself in a faraway port. Not surprisingly, Admiral Walker took many of the officers who had staffed the Bureau of Navigation with him when he left, offering them first pick of the "plum" assignments.[23]

From August to October the *Atlanta, Boston,* and *Chicago* moved from New York to Newport and back, completing their trials. The timetable for departure of the squadron was pushed back a few weeks when the *Boston* ran aground off of Newport on 3 August 1889.[24] Fortunately, the ship's new double-bottomed construction minimized the damage, and *Boston* was able to make her way slowly back to the New York Navy Yard, where she entered dry dock immediately. The setback with *Boston* notwithstanding, eventually the trials were complete and the ships returned to the navy yard, where they were fully manned and supplied. The squadron was made complete by the arrival of the gunboat *Yorktown,* which had been away on Navy Department tasking. On 18 November 1889, the four ships got under way together from New York. Secretary of the Navy Tracy and Admiral of the Navy Porter were among the notables who descended on the navy yard to see them off.[25]

Their first order of business had nothing to do with experimentation in fleet tactics. The squadron headed to Boston, where it joined in a maritime celebration taking place there.[26] Besides his interest in the development of fleet tactics, Walker understood the important public relations aspect of the White Squadron and took specific pains to make the ships accessible to the American public prior to taking them on their overseas cruise. He was pleased with the results of the time spent in Boston, where thousands of citizens had the opportunity to climb around on "their" new steel ships. According to Walker, "Probably not less than twenty thousand people . . . have been received on the Chicago." He went on to note that "from all sources are heard expressions of satisfaction that the United States is again taking position as a naval power, and I have been deeply impressed with the strength and sincerity of this feeling and the advantage which future naval legislation will probably derive from it."[27]

From Boston, the squadron set sail for Lisbon, Portugal.[28] Daniel Wicks points out that fleet tactical exercises could have taken place anywhere if they had been the only mission of the Squadron of Evolution, but there was more at stake here. A newly powerful U.S. Navy wanted the nations of Old Europe to be aware not only of the new ships it was fielding but also of the ability to deploy them across long distances.[29] It's worth pointing out, though, that a large segment of the public felt this way as well, not just the "navalists." Witness the words of the *New York Times* correspondent who wrote, "The presence of such a fleet for two years abroad will do more to secure respect to American travelers than a host of State Department documents well-worded but not backed up by a show of military force."[30]

After a two-week journey across the Atlantic, the squadron arrived at Lisbon.[31] Using his newly printed letterhead, which proudly proclaimed the "Flagship CHICAGO of the United States Squadron of Evolution," Admiral Walker reported that the three cruisers had weathered the crossing well, in spite of some heavy weather, but that the smaller *Yorktown* had become separated. He assumed that she had been forced to heave to by the weather and would rejoin the squadron in port in a couple of days, as indeed happened.[32]

The captain of *Yorktown,* French Ensor Chadwick, was uniquely qualified for duty with the new Squadron of Evolution. As one of the earliest proponents of the Office of Naval Intelligence, Chadwick had been posted to London, England, as the first U.S. naval attaché. While there he had corresponded regularly with then-commodore Walker, who was the chief of the Bureau of Navigation and Detail, about technological advancements in the various European navies.[33] He had also been instructed to produce a report for the Department of the Navy on the training systems of the British and French navies, which was forwarded to Congress in 1880.[34] Although he did not say much about ships operating in close order, Chadwick would have been more knowledgeable than any other officer in the U.S. Navy about the methods foreign navies used to conduct naval warfare.

In any event, Chadwick's *Yorktown* had indeed been forced to lie to in the bad weather, the seaworthiness of the little gunboat being further compromised by the parting of the ship's steering gear. After touching at Fusal for repairs, *Yorktown* rejoined the squadron on 23 December 1889.[35] In Portugal, Admiral Walker received the records of the U.S. naval force on the European Station from Commander McCalla of the *Enterprise,* who had been in temporary command of the station.[36] The squadron then proceeded to enjoy the hospitality of the Portuguese for the next ten days, putting to sea on 31 December with the expressed intention of "exercise[ing] . . . in squadron tactics under steam."[37]

This was Walker's first real opportunity to put the squadron through its paces . . . and he was not impressed. "The manner in which the Squadron got under way and took positions in column of ships," he wrote the secretary of the navy, "was unsatisfactory, showing that much practice in Squadron tactics is required to arrive at the necessary promptness and accuracy in handling the individual ships."[38] By 1 January 1890, the ships, having traveled in company for the past twenty-four hours, he had somewhat nicer things to say about the day's tactical work.[39]

Rear Admiral Gherardi and the North Atlantic Squadron, 1890

January 1890 found Rear Admiral Gherardi still preoccupied with affairs in Haiti and, in the manner of a traditional warrior-diplomat, spending much more time on diplomatic duties than training a squadron for fleet combat. After arriving in Port-au-Prince on 20 December 1889, his dispatch of 29 December made it clear that he was predicating his personal movements and those of his squadron on the arrival of the French minister to Haiti, with whom he hoped to have an opportunity to meet.[40] This was curious behavior, considering the fact that President Harrison had appointed a minister to Haiti, Fredrick Douglass. The resident minister should have taken care of meeting foreign dignitaries. The U.S. government, it seems, had little faith in Douglass' ability to conduct diplomacy and was counting on the presence of Rear Admiral Gherardi to make sure events played out in such a way to favor the interests of the United States.

Admiral Gherardi's flag was now flying from the little *Dolphin,* fresh from her 58,000-mile cruise around the world on steam power alone. *Dolphin* was the first of the ships of the New Steel Navy to be commissioned, and fittingly, she was assigned to the Home Squadron. At just over 1,400 tons, *Dolphin* was not designed for combat but to do utility work and deliver messages for station commanders. One of her secondary planned uses was as a flagship for a squadron commander and his staff, so Rear Admiral Gherardi's relocation from *Galena* represented the first opportunity to put that capability to the test. Her gunboat armament consisted of a single 6-inch breech-loading rifle and a pair of 6-pounder rapid-fire guns.[41] Small as she was, she represented the newest achievement in American shipbuilding, and the very fact that an admiral's flag was flying from *Dolphin*'s mast was a vindication of sorts for her builder, John Roach, by now dead for almost three years.[42]

Rear Admiral Gherardi did not stay on board *Dolphin* for long. As the newest and best of his ships, *Dolphin* was detailed to transport Minister Douglass to Santo Domingo to present his credentials to the government there. Not wanting to go himself—he was still anxiously monitoring the Haitian elections and the arrival of a new French minister—he shifted his flag to *Galena* on 9 January. It is evident that none of the uneasiness about hosting an African American that had marred Douglass' original transport to Haiti surfaced during this mission. Douglass' after-action report to the secretary of state was filled with praise for both the *Dolphin* and her crew.[43]

Gherardi returned to *Dolphin* when she arrived back in Haitian waters with Minister Douglass.[44] He then sent *Kearsarge* and *Galena* north under the command of the CO of the *Galena,* Commander Sumner, while the admiral and his flagship visited Havana, Cuba. Sumner and his charges were ordered to proceed to Matanzas, then Havana, Cuba, then to Key West, where they were to reprovision, recoal, and meet up with *Yantic.* While under way, the two ships were to exercise regularly in the "School of the Section," found in the Fleet Drill Book.[45] Gherardi kept his ships well drilled, but he was less interested in personally leading them than he was attending to political business on his station—namely, the ongoing negotiations for a U.S. naval base at the Môle St. Nicholas, a desirable harbor on the northern coast of Haiti. The Môle would provide an excellent vantage point for the U.S. Navy to keep an eye on the Caribbean and the approaches to any canal that might be built across the Central American isthmus.

Galena and *Kearsarge* arrived in Key West on 5 March, where they were met by *Yantic* two days later.[46] *Yantic,* which had been on special duty for the Bureau of Navigation, pulled into Key West on the seventh with the Longitude Party on board. This expedition, under the command of Lt. J. A. Norris, had left the United States in November 1889, charged with the telegraphic determination of longitudes in the West Indies and on the northern coast of South America. In this age of GPS, it can be hard to remember that in the late nineteenth century there were still major portions of the earth that were not charted accurately. As a rising world power, the United States became more and more interested in correcting these deficiencies. The geographic points surveyed say something about the United States' interests. The Longitude Party's first stop was Santiago de Cuba, followed by Môle St. Nicholas in Haiti. Upon arrival in Key West, the expedition unloaded their equipment and departed for Washington, D.C., arriving on 11 March 1890.[47] The continued presence of U.S. warships in Haitian waters caused problems for Minister Douglass. Within a month of his trip on board *Dolphin,* Douglass was writing his boss to complain about "speculation as to alleged designs of the United States upon the integrity of Haiti; speculation supported in part by the frequent appearance of United States vessels of war in Haitian waters."[48] This entreaty, and others like it, had little if any effect.

Meanwhile, the northern portion of the North Atlantic Squadron's area of operations was covered by *Petrel,* which was assigned to the squadron in June and cruised through the waters off the northeast coast. Rear Admiral Gherardi thought he would finally be able to gather his squadron together, without being distracted by Haitian politics, when he returned to Key West in March 1890. However, within a week he was summoned to Washington, D.C., to discuss the situation in Haiti. He delayed his departure from Key West long enough to observe his squadron take target practice, but eventually he was forced to leave the ships in the care of the senior officer present afloat (SOPA), Commander Sumner, and head north.[49] From there, he was forced to resort to ordering his squadron around by telegram.[50] Per Gherardi's orders, the squadron went to sea on 10 April and drilled for five days. Noticeably absent was *Yantic,* whose boilers were in such need of repair that she could not make the speed necessary to participate in squadron maneuvers. Upon Gherardi's return to Key West in April, *Yantic* was detailed to return to New York with a load of naval cadets and men whose enlistment terms were up. She was eventually transferred from the North Atlantic Squadron and put out of commission.[51]

Rear Admiral Gherardi did not have to wait long for a replacement ship to arrive. In May *Baltimore* arrived in Key West and Gherardi, acting in accordance with orders from the department, shifted his flag to the new steel cruiser.[52] She was part of the second generation of protected cruisers, built from plans originally drawn for the Spanish government that the Navy Department had purchased from Cramp's shipyards. At 4,600 tons and mounting a main battery of 6-inch and 8-inch breech-loading rifles, she was the first warship of the New Steel Navy to join the North Atlantic Squadron.[53] With the new flagship in place, the training program for the summer could begin. In keeping with the tradition that had been established over the last fifteen years, the squadron prepared to move north to conduct training during the hot and sickly summer months. Before departing Key West, a naval brigade of 350 men was put ashore, commanded by the executive officer of the *Kearsarge,* for practice in landings and naval infantry operations ashore.[54]

While the squadron trained, Rear Admiral Gherardi continued to be more concerned about conditions in Haiti. On 3 May 1890, he dispatched *Kearsarge* to run down to Port-au-Prince and "inquire about the condition of affairs in Haiti."[55] Political affairs by all rights should have been left to the representative of the State Department, while Gherardi concerned himself with the training of his combat unit. However, in keeping with the traditional role of naval officers as warrior-diplomats, Gherardi clearly felt that it was within his rights to inject his squadron's warships into Haitian politics. This slight did not go unnoticed by Minister Douglass, who complained in a letter to Secretary of State Blaine that "the presence of one of our national vessels in these waters is apt to attract general attention and to awaken curiosity and speculation."[56] Left unsaid (but clearly intended) was the observation that the captain of the *Kearsarge* was encroaching on his area of responsibility.

Fortunately, *Kearsarge* returned to Key West on 15 May with news that all was quiet in Haiti. Thus assured, Rear Admiral Gherardi was able to at last continue with his summer training program. The squadron sailed from Key West, proceeding to Charleston, South Carolina, where they paused off the coast of Jacksonville, Florida, for target practice on 28 May.[57] The target practice session featured the use of the new Fiske rangefinder, which had just been installed in *Baltimore.* The *Baltimore* fired her guns on both sides, steaming in large circles around a stationary target at ranges between eight hundred and two thousand yards. The rangefinder's inventor, as well as *Baltimore's* officers, was very pleased with the results. A 75 percent hit rate with the main battery was reported in the press; unprecedented if not entirely believable numbers.[58] The quality of U.S. gunnery seemed to be improving.

After stopping in at Port Royal, the squadron visited the port of Charleston from 5 to 8 June 1890. The four ships traveled in company, practicing tactical maneuvers along the way. From all evidence, this was the only practice of steam tactics undertaken by the squadron in 1890. After the Charleston port visit, the squadron proceeded to Portland, Maine, for a reunion of the Society of the Army of the Republic in July. In August President Harrison embarked on *Baltimore* for a trip to Boston, where the Grand Army of the Republic held a reunion. In company with *Baltimore* was *Philadelphia,* recently put in commission from the Cramp shipyards in her namesake city, Philadelphia. Officially designated Cruiser No. 4, she was a sister ship to *Baltimore,* sharing nearly the same hull and machinery but having a slightly different armament arrangement.[59] When the two ships returned to New York, Rear Admiral Gherardi shifted his flag to *Philadelphia* and *Baltimore* was assigned special duty from the Bureau of Navigation. She was tasked with returning the remains of Captain John Ericsson from New York to Stockholm, Sweden.[60] Of the remaining ships of the squadron, *Philadelphia, Petrel, Enterprise, Dolphin,* and the newly commissioned *Vesuvius* were together in New York, while *Kearsarge* was away on special assignment.[61] Without a doubt, much of the participation of the squadron in various commemorations throughout 1890 was done with an eye to showing the new *Baltimore* to a public supportive of continued expenditures on new ships.

In September Rear Admiral Gherardi detailed *Kearsarge* to proceed to Colón,[62] on the Isthmus of Panama, to "see that American interests [were] properly protected."[63] Her captain was Cdr. Horace Elmer. Once a young lieutenant on board *Colorado* during the Key West exercises of 1874, he was now a commander with his own ship—but still sailing alone to carry out national policy abroad. Upon arrival in Colón, Elmer found the situation quiet. In accordance with his orders from Gherardi, he got under way on 22 September for Greytown, promising to return by 1 October. The U.S. consul and the superintendent of the Panama Rail Road had asked for *Kearsarge's* presence on that date, which presumably was a payday, which might result in drunken unrest among the laborers in Colón.[64] Trouble came sooner than that. No sooner had *Kearsarge* got under way than a massive fire broke out downtown, burning most of the business district to the ground. The U.S. consul cabled frantically to Washington, D.C., begging for the return of the U.S. warship. Upon his arrival in Greytown, Elmer received a cable from the secretary of the navy directing him to return to Colón, which he did at once. It turned out that the fire was accidental, not the result of labor unrest, but that did little to stop Elmer from offering his less-than-complimentary observations about the citizens of the city. "With a city built of such material [wood frame buildings], warehouses full of such inflammable stores [liquor], a lazy idle and careless population, no efficient fire department, the first accidental fire was almost sure to result in its destruction." He proceeded in another report to blame the unrest of unemployed laborers in the city on "Jamaica negroes, ignorant, vicious, and troublesome."[65]

Across the Caribbean, Minister Douglass was still under fire. In October the State Department and Navy Department received a flurry of letters from William P. Clyde, owner of a steamship line that stood to profit greatly from concessions if the United States successfully gained access to Môle St. Nicholas. Apparently, the French were moving in on his business interests and Clyde felt that Minister Douglass was not doing enough about it. He was none too subtle in essentially calling Douglass incompetent. "After Admiral Gherardi left Haiti," according to Clyde's missive, "these people took advantage of Mr. Douglas's [sic] sympathy for the African Race and his lack of familiarity with the language and perhaps of diplomatic affairs to delay the carrying out of their pledges made to Admiral Gherardi as representing the United States government."[66] Despite his pleadings, Gherardi remained in New York and continued to plan for the squadron's winter activities.

Enterprise joined the squadron in New York in November and was promptly sent to Colón to relieve *Kearsarge,* which returned to Key West. Commander Converse reported things in Colón quiet, to the extent that by December his presence there was no longer deemed necessary, and *Enterprise* left Colón.[67] Later that month, Rear Admiral Gherardi released his winter plans for the squadron, which were eagerly reported on by the New York papers. The North Atlantic Squadron was to carry out independent cruising throughout the West Indies before rendezvousing at Key West in March 1891 for tactical exercises.[68] This was very much in keeping with the operation cycle of winter cruises in the Caribbean followed by work in company in northern latitudes in the summer that had been established for the squadron over the past decade. The *New York Times* congratulated Gherardi on his extensive plans for training in 1891, calling him "one of the most progressive officers in the navy."[69] It was curious praise, since the same paper in January had compared him unfavorably with the recently retired Stephen B. Luce, calling him a partisan of the "old school" and noting that fewer exercises and naval battalion landings had taken place under his leadership.[70] In fairness to Gherardi, the operations of the North Atlantic Squadron in 1890 show evidence of an organization in transition, exhibiting two different identities. The squadron now consisted of two steel ships, *Baltimore* and *Dolphin,* and two old wooden cruisers, *Kearsarge* and *Galena.* On the whole, it spent very little time conducting tactical exercises in 1890, yet it steamed in formation regularly and made port calls together.

Walker and the Squadron of Evolution, 1890

From 17 to 20 January, the Squadron of Evolution called at Cartagena, Spain, having first stopped at Gibraltar. Diplomatically, this was an important visit, as the situation in Cuba and the events of the 1870s were not far from anyone's mind. The Spanish authorities took great pains to show the American officers their facilities and warships under construction. The American delegation also toured the torpedo factory and school. In turn, Admiral Walker hosted a delegation of Spanish officers on board the four ships of the Squadron of Evolution, where they expressed, in the words of the vice consul, "their favorable opinions regarding the handsome construction, clean state, and the latest sea and war improvements and perfect order of the four ships."[71]

From Cartagena the squadron moved on to Port Mahon, Menorca, then Toulon and Villefranche, France, and finally to Spezia, Italy, arriving there on 3 March 1890. Here Walker sent off a long letter to the secretary of the navy, outlining an altercation he had had with Captain Howell of the *Atlanta*. Howell had allegedly not paid enough attention to the position of his ship in the squadron's formation, and eventually Walker—after repeated signals to *Atlanta* to improve her station keeping—relieved him. This action provoked a long letter of protest from Howell to the secretary of the navy, which Walker endorsed and forwarded with his explanation of the circumstances. In his words, "The Atlanta has repeatedly been very badly handled, not apparently through lack of seamanlike skill and judgment on the part of her commanding officer, but rather from an indifference to the tactical precision and appearance of the squadron and to the necessity for prompt and literal obedience to signals." Here we can see the clash of the old Navy and the new. An officer of the old guard, brought up under the old standards of professionalism, simply did not attach any importance to station keeping. He was used to being the master of his own vessel, reporting to the commodore for administrative matters, and occasionally sailing about somewhere in the vicinity of the commodore's ship, but not keeping a tight station at high speed. Walker noted that Howell "does not seem to appreciate the military requirement of his duty as the commanding officer of a cruiser in a *tactical squadron, and for the present service this appreciation is an officer's highest quality* [emphasis added]."[72] The work done in the intervening two months since his less-than-charitable remarks about the performance of his squadron in tactical drills had evidently paid off, as Walker noted in his report to the secretary of the navy that the "ships were much better handled, and the maneuvers were more satisfactory than ever before, showing that experience only is required to make the exercises all that is to be wished."[73]

Admiral Walker was looking for good weather, which the Adriatic had not provided, so he next took the squadron farther down the Mediterranean, calling at Corfu, Greece. The cruise of these four ships represented a sharp change from previous deployments. Dispatches from squadron commodores in previous years had mostly centered on the diplomatic and business functions of the U.S. naval mission. Commodores talked about the individual movement of ships, the ports they had spent time in, and the exchange of courtesies with the local authorities. Walker's dispatches contained those elements as well, but they also had a

new focus: his perceived mission to train his squadron. Walker chose to take the
squadron to Corfu not because the consul had requested it or because American
business interests were at risk, but because Corfu presented the best opportunities for the squadron to hold target practice and conduct landing exercises.[74]
In fact, although the consul to Greece desperately tried to get the squadron to
call at Piraeus (near Athens) so that the king of Greece could inspect the ships
and host the officers, Walker declined the invitation, giving the excuse that duty
required him to concentrate on taking advantage of Corfu's facilities for training the squadron.[75]

Upon leaving Corfu, the squadron was split up briefly. *Boston* and *Atlanta* proceeded to Messina, where they were docked and their bottoms were carefully
inspected and cleaned. *Chicago* and *Yorktown* went straight to Malta, arriving on
17 April, and the other two ships rejoined them a few days later. In Malta there
came another sign of the modernization of naval organization and another change
in the conception of the use of concentrated naval forces as national instrument
of power. A telegram arrived for Admiral Walker on 28 April instructing him to
keep his squadron in the Mediterranean until further notice. Walker had no way of
knowing that the desire of the secretary of the navy to keep the squadron together
and ready to respond as a unit had to do with the political situation in Brazil, where
the military had just overthrown fifty-eight years of rule by Emperor Dom Pedro
II and declared a republic. What he did know was that he had over a hundred men
whose term of enlistment was drawing to a close. An extension on deployment
of unknown duration would raise difficulties with these sailors. Additionally, the
squadron's stores had been purchased with the intent of returning to the United
States in June. Further time at sea would require the purchase of more supplies, and
Admiral Walker wrote to the secretary of the navy requesting information on the
nature of their orders to remain in the Mediterranean.[76] The letter does not address
openly Walker's prerogative as squadron commander to determine the deployment
and utilization of his assets, but that was almost certainly on his mind as he wrote it.
Walker was finding out first hand that an instrument of power such as a squadron
composed of multiple warships was a tempting tool for national policymakers to
use to influence political events ashore. U.S. Navy deployments were shifting from
the traditional role of showing the flag and protecting business interests to being a
display of national ability to project combat power.[77]

Admiral Walker departed Malta two days after the telegram arrived and
headed to Algiers. There another telegram was waiting for him informing him
that the squadron was to prepare for deployment to the coast of Brazil. He
left Algiers and proceeded to Gibraltar, where he procured the supplies he felt
necessary to carry out the orders. In his acknowledgment of his orders, Walker
could not help pointing out that had he been privy to the department's intentions, he could have focused more on the material condition of his ships, making

sure that they were ready for a cruise of uncertain duration to South America. As it was, he wanted to make sure that the secretary of the navy understood that he "placed before all other considerations not absolutely imperative, the training and tactical work of the Squadron of Evolution."[78] Perhaps Walker was simply hedging his bets in case there was a mechanical problem with one of his ships later, but his remarks still show a change in the concept of what the deployment of U.S. naval assets overseas should accomplish. Walker felt that it was his job to produce a combat-trained unit. Conducting diplomacy was secondary.

Once provisioned, the squadron left Gibraltar and touched at Tangier for one day, a quick reversion to the old practice of "showing the flag" in support of the new U.S. consul. It then proceeded to Madeira, arriving on 30 May 1890. Walker took advantage of the stop at Madeira to send another report to the secretary of the navy, bemoaning at length the fact that he had not been informed about the department's intentions for his squadron. Walker went so far as to intimate that the department did not trust him with sensitive information, in spite of his over forty years of faithful service.[79] After touching at Porto Grande, St. Vincent, Cape Verde, to communicate again with the department, *Atlanta, Boston,* and *Chicago* headed for Brazil, while French Chadwick and the little *Yorktown* were sent to New York with the squadron's short-time men, invalids, and prisoners.[80]

Admiral Walker's mood improved appreciably upon arrival in Brazil, where the squadron was greeted with great pomp and ceremony. The new military government was apparently anxious to reassure the United States as to its good intentions. Regardless of the reason, no one relished being paid attention to more than Commo. John G. Walker, and his report home was filled with details of the various honors paid to the squadron and him, personally.[81] Having fulfilled the State Department's intention to show goodwill to the new Brazilian government, the squadron headed back to New York. It arrived in July and spent the remainder of 1890 doing upkeep and various single-ship missions. Talk of combined exercises with the North Atlantic Squadron came and went with no action taken. Everyone recognized the unspoken tension between Gherardi and Walker. It would only get worse.[82]

The Two Squadrons Collide, 1891

By 3 January 1891, *Philadelphia* was ready to put to sea to lead the squadron in Rear Admiral Gherardi's plans for the upcoming year. Unfortunately, Gherardi was not ready. His wife's serious illness required him to ask for a delay in getting under way, which was granted.[83] *Dolphin* was at New York with *Philadelphia* and would get under way with her in January. The "dynamite cruiser" *Vesuvius* was at New York as well, although she was still conducting tests on her pneumatic guns and was not expected to be employed as a cruising vessel with the squadron.[84] *Kearsarge* was at Norfolk; Gherardi ordered her to get under way and meet him in Port-au-Prince. Commander Elmer pointed out that the amount of work necessary for *Kearsarge* would prevent her from getting under way as quickly as Gherardi would have liked, but he promised to meet him in Port-au-Prince as soon as possible.[85] *Petrel* was already in the West Indies, under orders to arrive at Port-au-Prince no later than 25 January,[86] and *Enterprise* was at Colón, Columbia. It was a pivotal year in the evolving identity of the North Atlantic Squadron. In 1891 old ships worked alongside ships of the New Steel Navy. Old deployment patterns and assumptions about responsibilities operated, at times spectacularly unsuccessfully, alongside new paradigms of squadron deployment and unit identity. At the center of that transition was an incident that took place in Port-au-Prince, Haiti, that drove to the heart of the question of the organizational identity of the two naval units simultaneously occupying the same geographical space.

Rear Adm. Bancroft Gherardi embarked in his new flagship was finally able to leave New York on 17 January.[87] He headed straight to Haiti to continue negotiating his government's earnest desire to have a naval base there. Upon his arrival on 25 January, he found *Petrel* waiting for him.[88] At 890 tons, the little gunboat was one of the smallest of the ships of the New Steel Navy, but she was heavily armed for her size and a versatile warship. The venerable *Kearsarge* joined the squadron on 5 February.[89] Rear Admiral Gherardi was under orders to treat with President Hyppolite of Haiti for the cession of Môle St. Nicholas. The negotiations soon ran into a snag, as it was discovered that Gherardi's commission had been signed by the secretary of the navy, not the president. For the Haitian officials, this would not do. They refused to talk further until a commission was delivered bearing the signature of President Harrison.[90] Gherardi had assumed that his position as an admiral in the U.S. Navy would provide enough diplomatic power to negotiate for a naval base, but finding that it did not, he took it upon himself to write home for a presidential commission. Rather than training his squadron for combat, Gherardi was acting the part of a warrior-diplomat of the old Navy.

Gherardi had asked for his special commission to be delivered by the Clyde steamer due in Port-au-Prince later that month, but the powers that be decided that it would be useful to have some more U.S. warships to provide a backdrop for the negotiations, and dispatched Acting Rear Adm. John G. Walker's Squadron of Evolution to personally deliver the commission.[91] This would turn out to be a fateful decision, not only for the negotiations themselves but also for the relationship of two of the Navy's most senior admirals.

From the start, the mission got off on the wrong foot with Acting Rear Admiral Walker. As usual, he was not given enough information far enough in advance about the department's wishes. When his orders (and, presumably, Rear Admiral Gherardi's special commission) arrived at Key West by registered mail on 13 April, the complaints started immediately. "If sent by open mail, I should have received them on the 10th, and if I had had any intimation of the duty required of me . . . the Squadron could have gone to sea on the morning of the 11th. I shall now have *considerable difficulty* [emphasis added] in getting to sea on the 15th."[92] For someone who was probably the most modern-thinking flag officer in terms of viewing his squadron as a single combat entity, Walker had a surprising amount of difficulty with urgent orders sent to him by telegram. As promised, he got under way on 15 April with *Chicago, Boston,* and *Yorktown*—the *Atlanta* remaining behind in Key West to have work done on her bottom. He arrived with his squadron in Port-au-Prince on 18 April 1891.[93]

The three white-painted steel ships steamed into the harbor and dropped anchor a mere three hundred yards from Rear Admiral Gherardi's flagship, without asking permission—a striking breach of both regulations and naval etiquette.[94] There are many versions of what happened next. Newspaper accounts emphasize the animosity between the two officers—an approach that obviously made more interesting reading. On the other hand, first-person accounts in print tend to have been cleaned up for publication. In any case, Rear Admiral Walker was summoned on board *Philadelphia,* where Rear Admiral Gherardi pointed out to him in no uncertain terms that he was the SOPA, that *Acting* Rear Admiral Walker was junior to him, and that it was expected that Walker's squadron would obey all signals from the North Atlantic Squadron flagship.[95]

It was a clash of the old school and the new. Walker felt that his squadron owed allegiance to no one but its commander, and that its commander owed allegiance to no one but the secretary of the navy. It was not the first time Walker had run into trouble on this point. He had had several arguments with various navy yard commandants prior to leaving for his European cruise, as well as with the Navy Department over making personnel decisions for ships in his squadron without consulting him.[96] It is unlikely that an officer as aware of his prerogatives as squadron commander as Walker was would not have been completely aware of the way in which he was snubbing Rear Admiral Gherardi.

Capt. Edward L. Beach Sr.'s memoirs relate that the two officers ended up having a drink in Gherardi's cabin and came out all smiles, the incident behind them.[97] This is perhaps fanciful, but in any event, Walker returned to his ship and lost no time in requesting permission to depart from the senior admiral. This permission was not granted for a week—for reasons Gherardi did not care to share with Walker. "I was detained there by Rear Admiral Gherardi," fulminated Walker in his report, "acting under the authority of the Navy Department, until 2:30 PM of the 24th instant, at which time I sailed for this port [Norfolk]."[98] Over the course of the next year, the two squadrons would ply the waters off the East Coast of the United States, carefully staying away from each other to avoid any further questions of seniority.

The collapse of talks with the Hyppolite government on 22 May 1891 showed, not for the last time, the shortcomings of "gunboat diplomacy." The proud Haitian government in the end simply told Gherardi and Douglass that it was not interested in ceding any of its land to the United States, and the Harrison administration was left with the option of taking Môle St. Nicholas by force or leaving. Unwilling to attack Haiti (and thereby risk alienating the African American vote, which had been instrumental in catapulting him into the presidency in 1888), the Republican Harrison was forced to take hat in hand and depart Port-au-Prince quietly. Minister Douglass resigned over the incident, remaining loyal to Harrison and assigning most of the blame to Blaine. The ham-handed attempts by Blaine, Tracy, and Gherardi to circumvent Douglass and use the Navy to carry out their designs on Caribbean hegemony had backfired.[99]

On his way back to the United States, Gherardi in *Philadelphia* stopped through Santo Domingo, where he met with President Heureaux and discussed the possibility of leasing the Samana Bay property that President Grant had been interested in twenty years earlier.[100] Although Heureaux was willing, his people were not, and nothing further came of the initiative. The North Atlantic Squadron finally returned to New York on 16 May. On the way home, Rear Admiral Gherardi's flag lieutenant, Lt. Allen G. Paul, died of what the papers called "brain fever," brought on by the excessive heat of Haiti's climate.[101] It was one final disappointing note to end what had been a professionally unrewarding cruise.[102]

With both squadrons now temporarily back in the states, one of the sillier naval episodes of the 1890s got into full swing: the wild machinations to ensure that the two admirals would not have to come in contact with each other. Walker's Squadron of Evolution was at Norfolk, refitting and having maintenance done. Gherardi's ships, with the exception of *Kearsarge,* which was still in the Caribbean, and *Vesuvius,* at Norfolk, were at the New York Navy Yard. It was well known that Walker wanted Gherardi's job, and equally well known that Gherardi had no intention of leaving his commander in chief billet until his prescribed tour was up.[103] While newspapers and seapower enthusiasts stirred up excitement about the possibility of combined-squadron fleet operations, both admirals demurred. Walker, who was in no hurry to subordinate his forces to Gherardi again, let the *Army and Navy Journal* know that he "[did] not think that there will be any joint maneuvers on the sound this summer."[104]

What Walker was up to was summer maneuvers of his own. Characteristically, he knew how to make a media event of it. On the first of July, *Chicago, Atlanta, Boston, Yorktown,* and *Newark* got under way from Norfolk, headed to Boston for the Fourth of July celebration. *Chicago* had engine problems immediately upon getting under way, so Walker transferred his flag to *Newark,* as they were on a schedule to be in Boston in time for the Forth.[105] *Newark* was one of the

Navy's newest ships, having been completed earlier that year. She was the first of the so-called second-generation cruisers to be authorized by Congress but the last to be completed due to complications during the design phase.[106] Improvements over the ABCD cruisers included triple-expansion engines, an increased steel protective deck, and better auxiliary machinery. She was well armed for a cruiser, with a main battery of twelve 6-inch breech-loading rifles, all mounted on the gun deck.[107]

Arrival in Boston was a gala affair and, again, Walker knew exactly how to get the most from an event. After the full-dress recognitions of the Fourth of July holiday, 5 July was designated as a general visiting day. Thousands flocked to see the ships of the White Squadron. On the sixth, official visits were exchanged with the governor of Massachusetts. One reporter noted that the state flag of Massachusetts flew from the foremast of *Newark* while honors were being rendered to the governor, testament to Walker's attention to the details of public relations.[108]

Another innovative event for the Squadron of Evolution was the embarkation of Massachusetts' battalion of naval militia for training and joint maneuvers. The militia members were drilled at the great guns and secondary battery. Subsequently the squadron got under way for live-fire exercises that were witnessed by the governor of Massachusetts, embarked with Rear Admiral Walker on board *Newark*. That night, members of the militia in small boats armed with simulated spar/attachable torpedoes made attacks on *Newark, Boston,* and *Atlanta,* which were simulating enemy ships in Boston Harbor. Newspapers credited the naval militia with successful attacks on two of the three ships. In his official report, Walker brushed off these claims, simply saying that "the attack, in all cases, had failed." The culmination of the exercises came on Friday, with a landing exercise on Deer Island. The sailors of the Squadron of Evolution formed one battalion, under the command of Commander French Ensor Chadwick, while the naval militiamen formed another. Together the two attacked a strongpoint on the island held by the squadron's Marines. After a series of balls and dinners the following day, the squadron sailed for New York. Two weeks later, the squadron repeated almost the exact same schedule of events with the naval militia of New York, conducting landing exercises on Fisher's Island off the coast of Connecticut. Again there were combined landings and naval infantry exercises, boat races, dinners, and celebratory parades.[109]

In all, Rear Admiral Walker predictably claimed success. "I regard the week spent in Boston as extremely interesting and valuable; instructive to the naval volunteers and encouraging to the officers and men of the regular service." In retrospect, the exercises probably did more to boost popular support for the New Steel Navy than make any actual preparations for a national emergency. The idea of militiamen piloting small boats with spar torpedoes to attach to enemy battleships (swinging quietly at anchor, no less) was laughable. However, the idea of having a trained core of civilians to man the ships the nation would have to produce in the event of a war had taken hold. "If war were to break out," remarked Walker at a reception given for the squadron in New London, Connecticut, "the great and pressing question at once arising would be as to the manning of our ships. . . . The Naval Reserve is a great need."[110]

Spending only a couple of weeks in New York, the Squadron of Evolution got under way on 12 August to cruise along the coast of New England. On board were several members of the Senate Naval Committee. Since the Boston/Fisher Island exercises, new ships had been added to the squadron. *Bennington* and *Concord* were sister ships of *Yorktown:* 1,700-ton gunboats that mounted six 6-inch breech-loading rifles and had hull openings for six torpedo tubes, although self-propelled torpedoes were not introduced into fleet use until 1894. These gunboats were often criticized for not packing enough firepower or protection to be effective fighters, but they proved to be the workhorses of the New Steel Navy—the "steam sloops" of the 1890s.[111] Thus enlarged, the squadron made its way up the coast, stopping at Newport, Boston, Bar Harbor, and New London, Connecticut. Everywhere along the way, the squadron's officers socialized with prominent citizens and were feted at dinners and balls. These events, while undoubtedly good for public opinion of the Navy (not that the citizens of New England ever had any trouble supporting a navy), also added fuel to the intense dislike that many both in and out of the Navy felt for John Grimes Walker. About this time, the *New York Times* began a series of articles exposing Walker's "pull" within the Navy Department and lamenting his cruises up and down the East Coast, entertaining and being entertained, while others did the daily work of overseas cruising.[112]

Meanwhile, Rear Admiral Gherardi's North Atlantic Squadron, without dinner parties or senators on board, slipped quietly from New York up to Bar Harbor, Maine, for exercises and target practice. As the Squadron of Evolution had grown, the North Atlantic Squadron had dwindled down, now comprising three ships— *Philadelphia, Petrel,* and *Enterprise,* with *Kearsarge* just back from the Caribbean at Norfolk for refitting.[113] From Bar Harbor the squadron moved to New London, Connecticut, departing quickly as Walker arrived with his Squadron of Evolution (and various senators).[114] Upon completion of their cruise to the north, the North Atlantic Squadron returned to New York, where Rear Admiral Gherardi busied himself making plans for winter operations in the Caribbean.[115] Gherardi's attention continued to focus on the Caribbean, and when his three warships sailed in November, they sailed separately and to different ports in the Caribbean. For the North Atlantic Squadron, operations in company were something that was done at a particular time of year for training purposes. The mission of the squadron and its commander in chief continued to be traditional cruising duty.

This may explain why the department turned to Rear Admiral Walker's Squadron of Evolution for the next political crisis, which occurred later in the year with Chile. There a civil war was under way pitting incumbent president J. M. Balmaceda and the Chilean army against the Chilean congress and navy, headed by Captain Jorge Montt. The congressional forces under Montt eventually prevailed, and by October 1891 it appeared for a while that the situation was going to quiet down. Things were calm enough for the commanding officer of *Baltimore,* the hapless Capt. Winfield Scott Schley, to make what can only be referred to in the kindest possible terms as an ill-considered decision to send his sailors into Valparaiso for liberty.[116] Predictably, violence broke out. In the resulting brawl, the Chilean police stood idly by while two sailors were killed and several others injured. Eventually the police dealt with the situation by arresting the Americans. Feelings ran hot in the United States, and when President Harrison went to Congress asking for a declaration of war, Tracy's Navy Department made immediate preparations.[117]

Conclusions

In the previous chapter we followed the North Atlantic Squadron as its identity began to slowly transition from an administrative organization to a combat unit. While exercises became more commonplace and the ships of the squadron quantitatively spent more time in company than they had in the previous decade, the focus of the North Atlantic Squadron continued to be traditional cruising, protecting U.S. commerce, and "showing the flag." This observation is confirmed by analyzing the operational movements of the North Atlantic Squadron and comparing them with those of a second squadron constituted in the home waters in 1889: the Squadron of Evolution. The Squadron of Evolution embodied everything the North Atlantic Squadron was still in the process of transitioning toward. Without ties to a specific geographic region and the political and diplomatic baggage such ties brought, the Squadron of Evolution was free to focus on what its commander repeatedly referred to in his correspondence as its primary duty: training together to operate as a combat unit. This focus resulted in a unit with an organizational identity that had as yet not been seen in the U.S. Navy. It was not the first time that the Navy had deployed its warships organized as a squadron; some of these had even developed a protean sense of organizational identity, such as "Preble's Boys" in the Mediterranean during the Barbary Wars, but none had been as consistently employed as a single combat unit, and thought of itself as such, as the Squadron of Evolution.[118] When its commander wanted to report its presence somewhere in the world, he sent a telegram that read simply: "squadron." Everyone in the Navy Department knew what that meant—that all four ships of the Squadron of Evolution had arrived. Those telegrams might be the best evidence of the development of the Squadron of Evolution's organizational identity.

The North Atlantic Squadron, on the other hand, although it had made great strides in operating as a single unit since the 1870s, still focused largely on solving foreign policy problems for the secretary of state. For all the evidence that Rear Admiral Gherardi took pains to exercise his squadron at naval tactics under steam, there is more that suggests that he was largely preoccupied with political matters, such as the fact that he was frequently called away from his squadron to Washington, D.C., for consultations. The alleged ineffectiveness of the U.S. minister to Haiti, Fredrick Douglass, meant that the State Department relied heavily on the admiral to carry out its policies on the island of Hispaniola.

Sadly, the relationship between the two squadron commanders precluded initiatives to exercise as a fleet. Perhaps it was inevitable that the two would clash, as they were part of a Navy whose command structure was largely unchanged since the 1840s. The personal bickering and habitual comparing of lineal numbers and privileges that had plagued the officer corps since the War of 1812 was still very much alive in the 1890s.[119] The senior officer corps was having a hard time catching up with the demands of a modern fleet, and the friction between the two squadrons in the summer of 1891 was proof of this. The *Army and Navy Journal* saw the need for better organization in 1891, writing in an editorial, "We should like to see some definite programme adopted for the use of the vessels of the American Navy which, in point of numbers, is growing daily, for its efficiency will continue to be seriously impaired until the officials cease working in the dark and adopt a definite naval policy."[120] As Assistant Navy Secretary James Soley put it in 1891,

> The old theory of squadron-cruising, in accordance with which a large force was maintained upon each of several foreign stations, where it lay for a great part of the time in port, and during the remainder cruised aimlessly about, is a thing of the past. Some force undoubtedly must continue to be maintained at certain points at all times, but the true place for a naval force in time of peace is in the waters that wash the shores of its own country. It is here that it should gain the practice that will enable it successfully to defend those shores when they are attacked.[121]

Whatever one thinks of the methods of Acting Rear Adm. John Grimes Walker, the Squadron of Evolution represented the future. While many government officials and naval officers pressed for this change in identity, the day-to-day operations of the North Atlantic Squadron over the next few years would demonstrate the fact that this change was happening very gradually.

4

★
The Limits of Ad Hoc Crisis
Response, 1892–1894

The Navy Department wanted to develop the ability of its officers and ships to fight in formation. Despite attempts throughout the 1880s and early 1890s to exercise the warships of the North Atlantic Squadron at fleet tactics under steam, the squadron had not yet developed a coherent identity. Contingency operations required the deployment of single warships from the squadron to safeguard overseas U.S. business interests, interfering with the ability of the squadron to develop a multiship fighting capability.

With the introduction of the ABCD ships in 1889, the Squadron of Evolution was formed to give a trained nucleus of officers experience at operating as a squadron. There were not yet enough of the new steel warships to distribute them throughout the Navy, so a considered decision was made to keep them concentrated. The North Atlantic Squadron, as has been seen, was largely left with cruising vessels to continue to carry out its mission during 1889–91. By 1892 new construction that had been authorized since 1883 resulted in enough matériel assets that the capability existed to keep a substantial concentration of modern steel warships together while deploying single vessels to handle diplomatic affairs and threats to business. The process of developing a fleet mentality, however, was not complete.

Rather than assign the new warships to the North Atlantic Squadron, giving it the ability to train tactically in the manner envisioned by Luce and, to a lesser extent, Walker, the new assets were put to use to further national objectives in foreign affairs and public relations. In late 1892, Rear Adm. Bancroft Gherardi was sent to California to lead a "Squadron for Special Service" around South America from the Pacific. The squadron's purpose was to deliver a message of goodwill to the major nations of Latin America and invite them to participate in the International Naval Review of 1893. The naval review, which was held in conjunction with the Chicago 1893 Columbian Exposition, saw the creation of the Naval Review Fleet, composed of two squadrons under two flag officers, with a fleet commander overseeing the formation. A fleet of this size and organization had not been constituted since the Key West maneuvers of 1874.

Historian Jan Rüger has identified naval reviews as a "theatre of power and identity that unfolded . . . in the imperial age."[1] A powerful navy was a way to not only influence world events but also build national identity—something that was critical to the post–Civil War United States.[2] The movements of the Squadron for Special Service and the participation of the Naval Review Fleet in the 1893 International Naval Review supported the Navy's public relations effort and afforded it extended opportunities to exercise operational and logistical challenges associated with maintaining large concentrations of warships. They did not, however, contribute to the development of the doctrine and tactics necessary for the North Atlantic Squadron to possess a multiship fighting capability. Naval reviews, public relations, and tactical maneuvers were not necessarily comparable strategic activities. The primary duty of the North Atlantic Squadron during 1892–94 continued to be the protection of commercial interests overseas, which the squadron accomplished with the deployment of single warships to trouble spots across its area of responsibility. While the Navy's new warships gained experience in multiship formations, those duties came at the expense of fleet tactical exercises. The process of constructing a new identity for the North Atlantic Squadron as a combat unit remained unfinished.

War Scare with Chile and Concentration in Montevideo, 1892

The fallout from the *Baltimore* incident in October 1891 pushed the United States closer and closer to war with Chile. The Navy Department began to consider its options for concentrating a force capable of contending with the Chilean navy. Unlike the confrontation with Spain over the *Virginius* incident in 1874, this was a fight the Navy Department felt ready to take on. Secretary of the Navy Benjamin F. Tracy was even accused of warmongering in an effort to show off his new warships. At least one newspaper editor at the time put forward the idea that action for the New Steel Navy would make Tracy a dark horse candidate for the 1892 presidential election: "A war with Chile would certainly develop one name as a compromise candidate between Blaine and Harrison. And that dark horse might be the horse marine, the alert, active, and accomplished . . . Benjamin F. Tracy, who would have more to do with the immediate making of war with Chile than all the rest of the cabinet combined."[3]

There is little evidence to support the editor's contention. When the Chilean revolution first broke out, Tracy went to great lengths to ensure that his commanders understood his wish to remain neutral and avoid trouble.[4] The deaths of two American servicemen may have lessened that resolve somewhat; still, he vehemently denied any desire for a war.[5] What is obvious are the immediate and aggressive steps Secretary Tracy took to place his New Steel Navy on a war footing. In early December, coded telegrams began flying from the Navy Department, moving ships, supplies, and coal toward South America.

The first unit to respond was, predictably, the remnant of Acting Rear Adm. John G. Walker's Squadron of Evolution. December 1891 found Walker's ships in Hampton Roads, as Secretary Tracy was doing his best to keep Walker away from Rear Admiral Gherardi of the North Atlantic Squadron, who was on the usual winter cruise in the West Indies at the time. The relationship between the two admirals since the difficulties in Haiti in the summer of 1891 had not improved noticeably, and since nobody in the department quite seemed to know what to do about the hostile relationship, the solution was simply to keep the two admirals as far apart as the East Coast would allow. This arrangement was, however, unworkable in an international crisis.

A coded telegram sent to Hampton Roads on 8 December 1891 initiated the formation of an ad hoc squadron to deal with the Chilean imbroglio. Walker was sent to the Dutch West Indies with *Chicago* and *Bennington*. The two ships were instructed to proceed to St. Thomas, where they arrived on 15 December to find a flurry of telegrams from the Navy Department concerning when and where to coal.[6] While the new Navy was enjoying matériel improvements in the form of new warships, getting them coaled and provisioned without a network of overseas bases was a chronic problem. The United States at this point was simply not prepared logistically to deploy concentrated combat squadrons across great distances. Secretary Tracy had been working nonstop since the crisis began to purchase and deliver coal to the U.S. warships headed south. The ad hoc nature of this process resulted in an unusually large number of telegrams issuing sometimes contradictory instructions to the fleet. The hectoring telegrams annoyed the self-confident Walker, who responded predictably. "I had no intention of coaling here [St. Thomas]," he retorted at one point. "I had not intended coming to this port, but came in obedience to Department's telegram of the 8th instant."[7]

The two ships were joined by *Atlanta* a day later, and all three got under way for Saint Lucia, where the department had American coal waiting. They arrived on 18 December and immediately went about the grueling business of coaling ship. Rear Admiral Walker was worried about yellow fever and for this reason did not want to stop in Brazil on his way to Montevideo. He thus ordered all three ships' bunkers filled to capacity. Additionally, the much larger (and less fuel efficient) *Chicago* and *Atlanta* took deck loads of coal.[8] Fresh water was another matter. Saint Lucia did not have the facilities to provide enough fresh water for the boilers of three modern warships in a timely fashion, so Walker made the decision to stop for a day at Barbados on the way south, where both the proper quantity of fresh water would be available as well as better facilities for pumping the water onto the warships. About this time, Assistant Secretary Soley made the mistake of sending Walker a fairly innocuous cable reminding him to reach Montevideo as soon as practicable.[9] This perceived slight inspired a four-page missive in response, as an enraged Walker railed against anyone who could question his abilities as a squadron commander. "Thus far, not an hour has been lost by this Squadron since leaving Hampton Roads," he fumed. He went on to find fault with the department's sailing instructions ("The call at that port was by order of the Department, and of the cause of the order I have no knowledge.") and the Bureau of Navigation and Detail ("The numerous changes of both officers and men, made at the last moment in the United States, have undoubtedly resulted in more or less inefficiency in the Engineers' force."), and he finished by petulantly noting that "the Department can reasonably believe that I have some knowledge of my profession. . . . If these ships do not reach Montevideo at the proper time it will be for reasons entirely beyond my control."[10] In the event, both *Atlanta* and *Bennington* had to stop through Bahia, Brazil, for coal, while Walker and *Chicago* obeyed department orders to press on to Montevideo, arriving on 10 January 1892.[11] He was joined by *Bennington* on 12 January and *Atlanta* on 15 January.[12]

Meanwhile, the decision had been made to supplement Walker's Squadron of Evolution with the rest of the modern warships available on the East Coast. On 24 December, Rear Admiral Gherardi, who was back in Haiti to assist the U.S. minister with managing possible unrest there, received a coded telegram ordering him to take *Philadelphia* and *Concord* and get under way for Barbados. There he was to coal and await further instructions.[13] *Yantic* was already in Montevideo, having left Hampton Roads in October 1891 to be transferred to the South Atlantic Squadron by way of Porto Grande, Cape Verde. She had arrived in Montevideo on 2 January 1892. Not finding any other U.S. vessels present, Lieutenant Commander Belden decided to proceed to Buenos Aires, Argentina, and report to Cdr. James M. Forsyth, the senior officer present in the area at the time. Forsyth was busy preparing *Tallapoosa,* one of the last of the wooden steam cruisers, to be sold for scrap.[14] In a somehow fitting commentary on the final demise of the old wooden cruising navy, *Tallapoosa* had literally rotted while assigned as a dispatch vessel on the South Atlantic Station, to the point that it was no longer safe to operate her. Her crew had to be sent home by merchant steamers, her supplies were salvaged by Rear Admiral Walker (who ordered *Essex* to proceed from Buenos Aires to Montevideo with whatever useful objects from *Tallapoosa* she could carry), and the hulk was ordered sold to the highest bidder at the dock in Buenos Aires.

Once settled in Montevideo, Admiral Walker made the diplomatic rounds, calling on the senior officers of the various navies represented in the port, the U.S. legation, and, on 22 January 1892, the president of Uruguay.[15] Meanwhile, supplies and coal ordered by the Navy Department continued to accumulate.[16] The plan was for the assembled ships at Montevideo to proceed around Cape Horn, through the Strait of Magellan, to Callao, Peru, where they would unite with the warships of the Pacific Fleet. To do this, they would require the services of a collier, which Rear Admiral Walker was ordered to arrange.[17]

Unfortunately, Walker was not able to get out of Montevideo with his three warships before Rear Admiral Gherardi arrived with his two. Gherardi's arrival automatically made him the senior officer present, and meant that Rear Admiral Walker would have to report to him. *Philadelphia* arrived at Montevideo on 6 February, followed a day later by *Concord*. Gherardi immediately went about making the formal calls on military and civilian officials that would be expected of the new U.S. senior officer present. It can be safely assumed that Walker, who took the integrity of his command so seriously and had just made all these same calls not three weeks earlier, was annoyed that his primacy on station, not to mention overall command of "his" ships, had been taken over by Gherardi. The *New York Times* reported that Walker stayed out of sight while Gherardi was in port, withdrawing to another anchorage to conduct target practice.[18] The *Times* notoriously disliked Walker, so its reporting must be accepted with caution. Walker did report to Gherardi on the movements of his warships while the two occupied the same station.[19] Throughout this period, in his official communications with Rear Admiral Walker, Rear Admiral Gherardi left absolutely no doubt about who was reporting to whom.[20]

On 25 January President Harrison went before Congress to ask for a declaration of war. Hours later the news that Chile had accepted unconditionally all U.S. demands reached Washington, D.C. The crisis was over, and it was perhaps as well for the harmony of the senior officer corps of the U.S. Navy. Within days of the president's message to Congress, Commodore Ramsey of the Bureau of Navigation was in touch with Admiral Gherardi, asking him how he wanted to be detailed after his current job.[21] Admiral Gherardi was subsequently recalled to home waters, to take up his post as commander in chief, North Atlantic Squadron for a few more months, prior to his relief.[22]

Walker stayed behind in his new assignment as commander in chief, South Atlantic Station. It did not take long for the new CinC to complain. Before Gherardi had even departed the station, Walker had already fired a five-page missive to the department. South America, it seemed, was unhealthy. There were not enough docking facilities available, and the bottoms of the steel warships had fouled rapidly. In any case, the "duties [on the South Atlantic Station] are inadequate to a Flag Officer's rank and position, and the expense is unnecessary."[23] It was apparent that Walker's real desire was Gherardi's position as the CinC of the North Atlantic Squadron, and he was making sure the department did not forget about him in the cruising backwaters of South America. Whatever his opinion of Latin America, he was still required to represent the U.S. flag there. The three ships departed Montevideo on 8 March and arrived the next day at Ensenada, Argentina. Walker and his personal staff boarded a train for Buenos Aires, where he was presented to the president.[24]

As happy as Walker was to have attention paid to him, his opinion of the station only worsened when he was made aware of an article in the *New York Herald* of 2 February 1892. It accused the sailors of the White Squadron with "riotous conduct" while on liberty in Montevideo. Walker fired off a three-page rebuttal to the department in which he denied that there had been any trouble, other than a couple of sailors who had been arrested for disorderly conduct. Whether or not that was the case, what is significant about Walker's letter is the closing, where he remarks that "it is due to the seamen of this Squadron and of the whole to give publicity to this authoritative denial of the false telegram enclosed."[25] Walker's salvation came not a moment too soon in his eyes. On 27 April, he was ordered back to northern waters.[26] *Chicago* left immediately, *Atlanta* following on 3 May 1892. Left on the South Atlantic Station were *Bennington*, *Yantic*, and *Essex,* under the command of *Bennington*'s captain as SOPA— precisely the way Walker wanted it. It seems that he still had some pull at the Navy Department after all.[27]

The Chilean incident and the ad hoc formation of a squadron to respond to it exposed the Janus-faced thinking at the Navy Department during this era. The standard operating practices of the Navy were still focused on cruising, showing the flag, and, most important, protecting property and commercial interests in foreign ports. However, at the first hint of conflict, the department concentrated its forces, anticipating some sort of naval battle utilizing fleet tactics under steam. These squadron concentrations were logistically and organizationally of an ad hoc nature. They did not contribute to the long-term construction of identity of the Navy's squadrons as combat units, nor were they as effective at developing a multiship fighting capability as they could have been. The exigencies of whatever international crisis was under way typically prevented the temporary fleet from carrying out any training to refine doctrine or tactics. Additionally, getting enough fuel to the warships in the absence of an infrastructure designed to support multiple ships was always a limitation, and the personal conflicts that resulted from flag officers being forced to work together without a clearly defined fleet organization hindered the Navy's effectiveness.

The North Atlantic Squadron: Operations, 1892

Rear Admiral Gherardi, in his flagship, *Philadelphia,* had departed Montevideo on 18 February 1892 in company with *Concord*. The two ships reached Barbados on 8 March, Gherardi remarking that *Concord* had spent most of the transit under sail to conserve coal, as the load of "Eureka" coal she had on board burned much less efficiently than the "Cardiff" coal that fueled the *Philadelphia*.[28] From Barbados, they touched at Havana and Matanzas before arriving at Key West on 8 April.[29] In Key West the old North Atlantic problem of piecemeal assignment of squadron assets immediately presented itself. On 9 April *Concord* was ordered to proceed up the Mississippi River to Memphis, Tennessee, and to be there by 12 May 1892. She was to support the gala festivities marking the opening of the Memphis Bridge between West Memphis, Arkansas, and Memphis, Tennessee.[30]

Concord's departure from Key West on 24 April for her trip up the Mississippi set off a wave of frantic telegrams from various civic associations along the way, begging the secretary of the navy to order *Concord* to stop at their municipality. Local vendors greedily eyed the money that the hundreds, if not thousands, of spectators who would flock to see one of the new steel ships would spend at their establishments.[31] While demonstrating the increased popularity of the Navy during the 1890s, *Concord's* goodwill trip up the Mississippi is evidence of the continued inability of the North Atlantic Squadron to build the kind of unit identity that Rear Admiral Luce had dreamed of for the squadron in 1886–87 and Rear Admiral Walker had made a reality for the Squadron of Evolution in 1889–91. In 1892 the North Atlantic Squadron still operated in a largely piecemeal fashion, with most of its warships away from the flag at any given time, covering contingencies and public relations events throughout the area of operations. *Kearsarge,* for example, remained behind in Haiti, while Admiral Gherardi and the rest of the squadron headed for Montevideo.

Newark, which had been detached from the Squadron of Evolution in late 1891 and prepared for cruise in Hampton Roads, was assigned to the North Atlantic Squadron in March 1892 and immediately ordered to La Guayra, Venezuela. *Newark* was Cruiser No. 1, the first ship of the 1885 authorization built and next in line after the ABCD ships. She had been commissioned in 1891 and assigned to the Squadron of Evolution for training prior to her assignment to the North Atlantic Station. At 4,000 tons, she was just smaller than *Chicago* but larger than *Atlanta* or *Boston.* Her main armament consisted of twelve 6-inch breech-loading rifles, and her new triple-expansion engines could drive her at a top speed of nineteen knots, four knots better than the best speed of the ABCD cruisers.

On 22 March 1892 Admiral Gherardi ordered *Newark* to leave St. Thomas and proceed to La Guayra, Venezuela, to "protect American interests as may be required."[32] *Newark* arrived two days later, on 24 March at La Guayra, a small port about twenty miles from the capital of Caracas. Capt. Silas Casey, *Newark's* commanding officer, immediately contacted the U.S. consul at that port and two days later journeyed inland to Caracas to call upon the U.S. minister to Venezuela. Casey found from his contacts with U.S. officials, as well as local informants, that a small insurrection was under way in Venezuela, but that violence was mostly taking place in the interior. The coastal cities and the capital were quiet, and Casey, in his report, remarked that although the insurrection was causing a general depression that might be bad for business, American lives and property were not in danger.[33] After a stay of another five weeks, Captain Casey concluded that his presence was no longer necessary and, in accordance with his orders from Gherardi, departed for Key West.[34]

Newark was able to spend just less than a week in Key West before getting under way for Savannah, Georgia.[35] There she joined Rear Admiral Gherardi in *Philadelphia* and *Kearsarge*, and *Vesuvius*, which had come down from New York. This concentration of warships for a goodwill tour of the East Coast represented all of the cruising vessels of the North Atlantic Squadron in May 1892, besides *Concord*, which was on her cruise up the Mississippi. As *Philadelphia*'s draft was too deep to allow her to enter the harbor, Gherardi transferred his flag to *Kearsarge*, which allowed him to tie up at the city docks and entertain prominent citizens of the town. *Vesuvius* tied up at the dock as well, and Gherardi reported that the two ships were visited by "thousands": "The visit . . . cannot fail to be of benefit, by means of the increased interest in the Navy that has been created."[36] Especially popular with the visiting crowds was the former enemy *Kearsarge*, which had won fame by sinking the Confederate raider *Alabama* so many years earlier. *Philadelphia* had to remain several miles out at sea, but even so, Gherardi reported that many steamers and tugs had brought visitors out to see her.[37] Rear Admiral Gherardi and his officers were feted at special dinners, and the Marine guards of the four ships marched in a parade through town.[38]

Altogether the squadron spent eleven days in Savannah, departing on 18 May. *Philadelphia*, *Newark*, and *Vesuvius* proceeded to Charleston, South Carolina, where the local chamber of commerce had invited the squadron to stop on their way north. Rear Admiral Gherardi's acceptance of this invitation again shows his mindfulness of the public image of both the Navy and his squadron in the port cities in his jurisdiction.[39] *Kearsarge* was sent to Port Royal for coal, with orders to rejoin the squadron when able.[40] At Charleston the squadron was once again feted ashore, and it was visited by what Gherardi referred to as "large numbers of the people."[41] They sailed for Annapolis on the evening of 21 May, arriving in time to take part in the Naval Academy's "June Week" graduation celebrations.[42] After the festivities at Annapolis, the squadron withdrew to Hampton Roads to await the arrival of *Concord*, fresh from her goodwill trip up the Mississippi to Memphis. From there, *Philadelphia*, *Concord*, and *Vesuvius* returned to New York, while *Newark* stayed behind for work at the Norfolk Navy Yard and *Kearsarge* was tasked with towing the monitor *Passaic* to Boston.[43]

Gherardi only spent three weeks in New York before getting under way for drills and target practice. The destination was Gardiner's Bay, Long Island, where the North Atlantic Squadron routinely held gunnery exercises. Gherardi sent *Miantonomoh* ahead ten days early, as her slow speed would have held back the rest of the squadron had she been in company. After target practice, the squadron reunited at New London, Connecticut, where the Naval Station had facilities for small arms practice.[44] At New London there was some discussion of a trip up the St. Lawrence River to the major port cities of Canada, but this did not materialize.[45] The squadron left New London instead for Bar Harbor, Maine, on 6 August to take part in festivities there.[46] It was joined at Bar Harbor by *Dolphin,* which embarked the secretary of the navy to witness a naval review that concentrated the available warships on the East Coast.[47] Four hundred sailors were landed to march in a parade through town, while Secretary of the Navy Tracy entertained such luminaries as J. P. Morgan at a dinner given on board *Dolphin.* From Bar Harbor, the squadron proceeded to Gloucester, Massachusetts, where it was met by *Miantonomoh.*[48] Again, the duty assigned them was to participate in a celebration of the 250th anniversary of the founding of Gloucester.[49]

The warships of the North Atlantic Squadron spent a good deal of time together in the summer of 1892. In this way, they continued to develop their unit identity. What was missing was any mention—official or otherwise—of tactical exercises or formation work during the months from June to September. The concept of the squadron as a tactical unit had not fully taken hold yet, and while the daily work of formation steaming was useful from developing a concept of multiship operations, it did not contribute to the doctrine and tactics necessary to possess a multiship fighting capability. At least one contemporary observer was mystified by this. Retired captain Edward L. Beach Sr. noted in his memoirs that the "*Philadelphia* was always active with drills, but seemed to me to steam about purposelessly. Instead of following a carefully arranged program, as would be the case today [the 1930s], we were able to attend flower shows and carnivals from Maine to Florida. We seldom cruised in squadron formation."[50]

Rear Admiral Walker, on the other hand, made a point of keeping his squadron together. His official communications and telegrams with the Navy Department often refer simply to the "Squadron" when reporting their movements, understanding that the department would know which ships were his and would assume that they were together.[51] As previously noted, Walker departed from the South American station in April. The station was without a CinC for a few months, until Rear Adm. A. E. K. Benham raised his flag on board his new flagship, *Newark,* on 25 June 1892. *Newark* had returned to Norfolk and was fitting out for her cruise at the Norfolk Navy Yard at the time. Benham and *Newark* left the yard in July, cruising across the Atlantic to Spain to represent the United States at the celebration of the four hundredth anniversary of Christopher Columbus' departure for the New World. Their duties in connection with the multination commemoration took them next to France, then to Italy, before finally passing Gibraltar on the way to their new station in 1893. Walker's opinion that the presence of a flag officer on the South American Station was not necessary was at least partially validated, as there was no CinC in the region for almost an entire year.

Speculation ran rampant in New York City and Washington, D.C., as to what the new flag officer assignments would be at this time. Both Gherardi and Walker were completing sea tours and were due to be rotated ashore, if not retired.[52] In a series of surprising moves, however, both senior officers managed to stay at sea. Rear Admiral Walker took command of the North Atlantic Squadron on 10 September 1892. Gherardi was given command of something to be called the Squadron for Special Service. It is an interesting juxtaposition of duties, as Gherardi would now have a squadron formed for the express purpose of staying together and operating as a unit, while Walker–who had spent the last three years with just such a unit, whose identity and image he had carefully cultivated, would now command a unit that had not yet fully developed an identity, and whose members were constantly changing or being sent on individual missions by the Navy Department.

Walker as Commander in Chief

It did not take the new commander in chief long to chafe at his situation. The week before he took command, he was sent with *Chicago* to the Naval War College to participate in the maneuvers associated with the fall course. Unfortunately, *Chicago* was the only ship assigned by the department to carry out the maneuvers, a situation Walker found ridiculous. In one of his characteristic four-page litanies of dissatisfaction, Walker complained, "I have absolutely no knowledge of what I am expected to do here with my command." He went on to note that more ships were necessary to carry out the turning radius and speed trials the Naval War College class wished to study: "If anything is done here, by my ships, while I am in command, I mean that it shall be well done; and therefore, write this to urge that the ships coming here may be sent at once."[53] It does not seem that he was upset about working with the Naval War College. Indeed, much of his correspondence with Rear Admiral Luce five years earlier when he was the chief of the Bureau of Navigation was positive about things he could do to help the Naval War College project.[54] Instead, it simply seems that Walker was annoyed because he did not have a formation of ships to direct. He even signed himself the commander in chief of the "Squadron of Evolution," although the squadron at that point consisted only of *Chicago*. After the change of command on 10 September, Rear Admiral Gherardi's old flagship, *Philadelphia*, reported to Walker for duty, and the two ships (he was allowed to keep *Chicago*) prepared for sea.[55]

Gherardi and the Squadron for Special Service

In the meanwhile, Rear Admiral Gherardi left New York for San Francisco in mid-September, accompanied by his flag secretary, Lt. William Potter, and flag lieutenant, Lt. Ridgley Hunt. The party stopped at Chicago on the way to inspect the construction under way for the world's fair to be held the next year. Once in San Francisco, Gherardi was to raise his flag on the *Baltimore*. Together with *Charleston* and *San Francisco*, the three ships would begin a special diplomatic cruise down the western coast of Central and South America, stopping first at San Diego, California, to take part in the commemoration of the 350th anniversary of Cabrillo's exploration of the coast of California.[56]

The Squadron for Special Service can be seen as an attempt by the U.S. government to strike a conciliatory tone in Latin America after the fiasco with Chile in 1891–92. While traveling down the western coast of South America, the squadron was slated to stop at each of the major Latin American ports, where Rear Admiral Gherardi, himself no stranger to the diplomatic side of his job, would personally deliver an invitation from the president of the United States for the nation to send representatives to the naval review scheduled for spring 1893 in New York City.[57]

Gherardi was insistent on having the latest and best equipment for his squadron, knowing not only that they were going to be spending a lot of time in formation, but that other navies around the world were going to be observing them. One of his first acts upon assuming command of the Squadron for Special Service was to ask the Navy Department to ensure that his ships were equipped with the latest Ardois night signaling capability.[58] The Ardois system was a series of five groups of two lamps, red and white, which were hung from stays in the rigging and operated by a keyboard on the bridge. By illuminating different five-digit combinations of red and white, orders could be transmitted to the rest of the fleet. The system could only be utilized by ships equipped with auxiliary electric power. Fortunately, all the vessels in the Squadron for Special Service were so equipped.[59] Gherardi had not shown this level of interest in communications during his assignment as the North Atlantic Squadron commander in chief. In fact, an analysis of 282 Gherardi communications, covering all of his afloat flag assignments, reveals that this letter was the first time he addressed an issue of signal communication and fleet maneuvering. His concern in this area was related to the public relations aspect of his mission and the appearance of power. The fact that the Squadron for Special Service had not developed any substantive fighting capability through rigorous exercise was secondary to the appearance of such a capability.

Baltimore and *Charleston* left Mare Island, San Francisco, on 25 September 1892 for the three-day voyage down the California coast to San Diego, where the celebration to honor the anniversary of Cabrillo's exploration was under way. The two ships arrived on 28 September. *Charleston*'s sailors were put ashore to march in a parade, and Gherardi received the governor of California and other notables onboard *Baltimore*. He also took the opportunity to pay an official visit to *Charleston*, which had recently joined his flag. He pronounced himself "pleased with the neat and efficient appearance of both vessel and personnel." Thousands of civilians visited the two ships as they took part in the commemoration.

It was not all pleasure for the two ships' companies, however, as Gherardi ordered first *Charleston* then the flagship *Baltimore* out into the bay to complete their quarterly target practice. On 5 October, the Naval Reserve of San Diego reported onboard *Charleston* for a day of drill.[60] Then it was off to Mazatlan, Mexico, arriving on 15 October 1892. On the way down from San Diego, Gherardi reported that the "ships of the Squadron were exercised at fleet tactics and signaling, especially night signaling [*sic*] by the Morse code with the steam whistles." Gherardi's report does not give any detail as to the tactics practiced, but the implication is that the drills involved the formations described in Parker's *Fleet Tactics Under Steam*. Gherardi also forwarded the department a copy of the "Routine and Instructions" established by the flag for the squadron.[61] The squadron called next at Acapulco, Mexico, and then continued to San Jose de Guatemala, where the admiral traveled to the capital of Guatemala on a special train to call upon the Guatemalan president.[62] This process was repeated in Peru a few weeks later. A state dinner was given in the Peruvian capital of Lima for the officers of the squadron.[63] In Peru, Gherardi got down to some of the most important business of the voyage of the Squadron for Special Service. On 3 December 1892, he and the U.S. minister to Peru made an official call on the Chilean minister. One year out from the very real threat of war between the two countries, Gherardi was feeling out the reception *Baltimore* could expect in Valparaiso, their next destination. The Chilean minister returned Gherardi's call on board *Baltimore* the next day and was received with honors and a fifteen-gun salute. The goodwill visit to Valparaiso would proceed as planned.[64] In Callo, Peru, *Yorktown* joined the squadron.[65]

Rear Admiral Gherardi continued to be concerned about the unity of his command even as he was carrying out his diplomatic mission. While in Peru, he wrote the department to suggest that any ships that were going to be assigned to the upcoming Naval Review Fleet, of which he was going to be the commander in chief, be scheduled to meet his Squadron for Special Service in Barbados, so that they could practice formation steaming prior to arriving at Hampton Roads or New York. "It is absolutely essential that the ships should be able to maintain position accurately in column, both on a straight course and when changing direction, and be able to act together in getting underway," he stated. He went on to note that these seemingly simple evolutions often were not simple at all: "My experience since leaving San Francisco is a fresh illustration of this fact."[66] Gherardi's major concern, both for the Squadron for Special Service and the upcoming naval review, was the smart appearance of his command.

Gherardi exercised his four ships at fleet tactics on the way to Valparaiso.[67] The squadron anchored at Valparaiso on the sixteenth of December. Salutes and official visits were exchanged, with both sides eager to display the utmost civility to each other. The *intendente* of Valparaiso came on board to express his hope that Gherardi would give the men of the squadron liberty in his city and offered to help facilitate that. Gherardi politely declined, pointing out that he was in a hurry to get home and begin his preparations for the naval review. What he failed to point out was that he had received a frantic coded telegram from the Navy Department while in Peru: "Do not give your men liberty at Valparaiso. Ramsey."[68] Gherardi was only too happy to oblige. He went ashore, however, to make his official calls, and he pronounced the civilian populace "peacefully inclined."[69] On 20 December Gherardi and his staff boarded a special train that had been provided for them by the Chilean government for the trip to Santiago. Gherardi's report describes the visit to the capital as proper though not enthusiastic. He noted that they were not welcomed at the train station by anyone other than the U.S. minister. His audience with the Chilean republic's president, Admiral Monett, was largely perfunctory. Gherardi informed Monett that the president of the United States had personally asked him to bring four of their newest warships to Valparaiso to deliver an invitation to take part in the naval review scheduled for the following year. Whether the Chilean president accepted this as anything other than a thinly veiled threat is hard to say, but he politely demurred, saying that it was too expensive for the young government to attempt to send one of its ships to New York. An invitation to visit the squadron was similarly declined, and after both men expressed their satisfaction that the two nations would recover their previous good relations, Gherardi was dismissed. "I was not encouraged to delay my departure" was the slightly more diplomatic way he described it in his report.[70] There would be no state dinners in Chile. The squadron's next stop was Montevideo, the scene of the fleet gathering eleven months earlier.

Walker, the North Atlantic Squadron, and Unrest in Venezuela

Rear Admiral Walker, meanwhile, had to satisfy himself with the mundane duties of the North Atlantic Squadron, which included minor crisis management across the Caribbean region, safeguarding U.S. business interests. The first exigency was unrest in Venezuela, where a faction of the Venezuelan congress was fighting against an incumbent president who refused to cede power. There were reports of violence against the U.S. minister there, as well as molestations of U.S. citizens and property. *Concord* was the first North Atlantic Squadron warship to respond, leaving from St. Thomas, Dutch West Indies, and arriving at La Guayra on 14 September 1892. There, Commander White, *Concord's* commanding officer, made an initial assessment of the situation.[71] Meanwhile, the Navy Department dispatched additional forces. *Kearsarge* departed Port-au-Prince, Haiti, on 9 September, and after leaving orders for *Philadelphia* to represent the squadron at the Annual Encampment of the Naval Veterans of the United States in Baltimore, Walker himself left New York on board *Chicago* on 11 September.[72]

When Walker arrived at La Guayra on 19 September, he found *Kearsarge* and the gunboat *Concord* in port. After conferring with their commanding officers,

Commander Crowninshield and Commander White, and meeting with U.S. consular officials ashore, he came to the conclusion that the situation in Venezuela regarding U.S. citizens and property was largely stable. *Concord* was dispatched to Puerto Caballo, a port city near the border with Colombia, about a day's sail from La Guayra, with orders to prevent interference with the movements of the U.S. mail steamers that called there and to report back anything of interest to Walker.[73] *Kearsarge* was sent to Coro, Venezuela, with similar orders.[74] The Venezuelan congressional forces under General Crespo enjoyed substantial support from the population at large and were widely expected to prevail in the civil war. When they did, it was assumed that the situation would quickly revert to peace. Having said that, Walker, reporting to the department on the situation, went on to complain about the length of time since *Chicago* had been in dry dock and to note that the European naval forces had all assessed the situation as stable and withdrawn from La Guayra. He then requested to be recalled to New York as soon as possible.[75] The Navy Department answered with a coded telegram on 8 October, ordering a reluctant Walker to remain in Venezuelan waters until further notice.

The fact that he was unhappy with his posting in Central America may account for Walker's worse-than-usual temper during this time. On 29 September, Commander Crowninshield of *Kearsarge* was the recipient of a scathing letter in which Walker upbraided him for the sloppy performance of his boat crews during a boat drill controlled by the flagship. "You will require each line officer under your command to read the instructions for sail and spar drill and for boat drills, to make a copy of them and to report to you that he understands them," he wrote.[76] *Kearsarge,* as an older wooden sloop, was accustomed to single-ship operations and not used to operating under the eyes of the flagship.

Rear Admiral Walker eventually was able to depart Venezuelan waters. On 19 October, after spending the opening portion of his report of that date recounting the political news from Venezuela, including the news that the congressional forces had taken the capital and the port town of La Guayra, Walker once again opined that there were too many U.S. warships present and informed the department that he was going to St. Thomas for coal and requested further orders from there.[77] No more coded telegrams were forthcoming from the department, and Walker did as he promised, arriving at St. Thomas, Dutch West Indies, on 28 October 1892. Before he left, he dispersed his warships to investigate various incidents that concerned U.S. businessmen, taking such action to safeguard U.S. lives and property as the situation(s) might warrant.

The International Naval Review, 1893

Rear Admiral Gherardi reported the arrival of his Squadron for Special Service at Montevideo, Uruguay, on 9 January 1893. The squadron had left Valparaiso, Chile, on 24 December 1892, passing through the Strait of Magellan on its trip to Uruguay. After a stay of two weeks in Montevideo, during which time visits were exchanged with the president of Uruguay, the squadron got under way for Barbados.[78] The time spent in Montevideo gave Gherardi an opportunity to complain to the Navy Department about the practice of sending communications to his ships without first passing them through the flagship. This was an issue that had vexed Rear Admiral Walker and his Squadron of Evolution. Gherardi had not dealt with issues of chain of command often, as the North Atlantic Squadron during his tenure was rarely together in one place for long enough for it to become a problem.[79]

The squadron left Montevideo on 23 January and proceeded to Barbados, where they arrived on the morning of 12 February. Gherardi was forced to disperse his squadron throughout the West Indies, as there was not a port that had the facilities to coal all four ships expeditiously. *San Francisco* and *Yorktown* went to St. Lucia, *Charleston* stayed at Barbados, and *Baltimore* headed to St. Thomas. Much like the fleet concentration in Montevideo the previous year, the fact that Gherardi had to break up the squadron in order to coal and replenish the ships shows that the logistical resources of the Navy had not yet matched the desire to deploy multiship squadrons on a regular basis.[80] Coal consumption was a subject of constant concerns for the bureaus of the Navy Department, and soon after arriving back at New York, Gherardi was asked to submit a report showing by ship the amount of coal bought at each port and the price paid.[81]

Gherardi also reported to the Navy Department on the feasibility of sending Morse code or Myer code signals using the ships' steam whistles.[82] The Signal Office was constantly on the lookout for better ways for ships to communicate at night and in bad weather. On the whole Gherardi thought that the modified Myer code worked better with the steam whistles, although both methods were slow, requiring as much as twenty-seven seconds per word.[83] While the Squadron for Special Service was not directly adding to the Navy's ability to fight in multiship formations, it was continuing the process by providing a venue to test such fleet essentials as effective signaling. The squadron left Barbados on 18 February, bound for Hampton Roads, the home waters of the East Coast, and their next assignment at the Columbian Exposition of 1893.[84]

The mid- to late nineteenth century witnessed an explosion in the popularity of large, international fairs and expositions. A series of world's fairs held in London and Paris from the 1850s to the 1880s proved immensely successful as celebrations of European progress and wealth. By the 1890s, the United States was ready to host a lavish debut onto the world stage. The four hundredth anniversary of Columbus' discovery of the New World proved to be a perfect opportunity. The 1893 Chicago World's Columbian Exposition, which ran from 1 May to 30 October 1893, is most famous in popular memory for bringing us Cracker Jack and the Ferris wheel. More serious historians have treated the underlying themes of race, gender, and class that underscored the opulent undertaking.[85] Rarely, though, does one run across mention of one of the major components of the celebration, because it took place some one thousand miles from the fair itself, in the waters off New York City.

The International Naval Review was scheduled to be held in conjunction with the Columbian Exposition in Chicago. Congress had appropriated $250,000 for this purpose, and it expected that every available warship of the New Steel Navy would be represented. On 1 March 1893, the Squadron for Special Service, the North Atlantic Squadron, and the South Atlantic Squadron were disestablished. In their place, created by the secretary of the navy through Special Order No. 2, was the Naval Review Fleet. Commanded by Rear Admiral Gherardi, the Naval Review Fleet consisted of the First Squadron, under the South Atlantic CinC, Rear Adm. A. E. K. Benham, in his flagship *Newark,* and the Second Squadron, with Rear Adm. J. G. Walker as the commander in his flagship *Chicago.* The Patrol Division was also established, with Capt. Frederick Rodgers in command. This fleet was the largest concentration of U.S. naval power since the Key West maneuvers of 1874, its establishment marking the first time since the Civil War that an attempt was made—although largely administrative and not tactical—to bring ships of various types and missions together under a single commander.

Anxious to be back on board "his" flagship after his trip with the Squadron for Special Service, Gherardi shifted his flag to *Philadelphia* on 20 March. Although the two vessels were sister ships, built by the same shipyard, *Philadelphia's* quarters and "certain other arrangements" were preferred by Gherardi for use as his flagship.[86] From the beginning Gherardi made it clear that he expected the two squadrons and all of their assigned ships to act as a unified command while the Naval Review Fleet was in existence. A "Routine and Instructions" was published in booklet form and copies were sent to every ship. On 1 March, Rear Admiral Walker immediately and properly reported to Gherardi in New York. This marked at least the fourth time in the last three years that Walker had been placed under the direct authority of Rear Admiral Gherardi. For someone who was so concerned about the image of his squadron and had coveted the North Atlantic CinC position for so long, to have communications to and from the Navy Department now have to pass through Rear Admiral Gherardi, subject to his endorsement, must have exasperated Walker. The obvious glee with which the New York newspapers speculated on Walker's humiliation did not help matters.[87] Throughout the Naval Review Fleet correspondence, virtually every communication that had to do with Rear Admiral Walker had a slightly negative tone about it. On 24 March Walker complained to the Navy Department that Rear Admiral Gherardi's staff had not provided his ship with copies of the new "Wig-Wag" signal book. Gherardi's chief of staff put an exasperated-sounding endorsement on the complaint, noting that *Chicago* had been provided with twenty-five copies of the new signal book but that the flagship would be happy to send over another thirty copies, addressed specifically to Rear Admiral Walker.[88] On 29 March an annoyed Gherardi responded to a Navy Department query about landing facilities at Norfolk, stating that he had asked Walker that very question already and was still waiting for the reply. "As soon as I reach Hampton Roads, the existing landing facilities will be carefully examined." It does not take much to read between the lines the fact that Walker had simply ignored Gherardi's instructions to carry out the task at Norfolk."[89]

Walker's North Atlantic Squadron was, characteristically, not concentrated but spread throughout the area of operations. *Atlanta* was at Key West, having recently returned from relieving *Kearsarge* in Haitian waters; *Concord* was in Norfolk, just back from her mission to Colón, Colombia; and *Miantonomoh* was in New York, from whence she rarely ventured due to her slow speed and poor sea-keeping qualities. The North Atlantic Squadron was rounded out by the dynamite cruiser *Vesuvius,* which was on her way back to New York from a stay at the Torpedo Station at Newport, Rhode Island, to load her distinctive projectiles.[90] From the South Atlantic Squadron, Rear Admiral Benham was returning from his mission to Europe, where he had taken part in several commemorations of Christopher Columbus' European connections before leaving Cadiz, Spain, with *Newark* and *Bennington*. The two ships were towing across the Atlantic two of the three replicas of Columbus' caravels, *Nina, Pinta,* and *Santa Maria,* which the Spanish government had donated to the exposition.

Bringing this diverse collection of vessels together as one unit was a problem that continued to occupy much of Rear Admiral Gherardi's time. One of the first things that he observed was that there was no standard paint scheme for the New Steel Navy, so he ordered that the masts and stacks of all ships under his command be painted to match the flagship for "uniformity in appearance." There is no evidence that Gherardi was concerned about this as CinC of the North Atlantic Squadron. Probably, his ships were never together long enough for the issue to become apparent to him, but with twelve ships guaranteed to be together for the review, the problem had to be addressed. Gherardi's order caused objections from the bureau chiefs, who felt that they were responsible for the painting of ships, but he got his way and the ships were repainted.

Rear Admiral Walker departed New York for Hampton Roads on 15 March in his flagship *Chicago*. Already there, shuttling between the Norfolk Navy Yard and anchorage in Hampton Roads were the *Newark, San Francisco, Charleston, Atlanta, Concord,* and *Dolphin*. The *Bennington,* which had been engaged in towing the replica caravel *Pinta* across the Atlantic from Spain, arrived on 26 March.[91] On the thirtieth, Rear Admiral Gherardi left New York in *Philadelphia,* with *Baltimore, Vesuvius, Yorktown,* and *Cushing* in company. The warships arrived at Hampton Roads the next day.[92] Rear Admiral Gherardi immediately began shaping his command into a fleet. The correspondence at this time shows a leader who was aware of rivalry among the flag officers and taking extra pains to establish a clear chain of command. Within hours of his arrival in Hampton Roads, he fired off a curt note to Rear Admiral Benham, who was already in Norfolk with *Newark:* "I can only explain your failure to report to me since your arrival by the supposition that you have not received [General Order No. 1, establishing the Naval Review Fleet].[93]

His fleet's chain of command and uniformity in appearance were two things on the admiral's mind, but signaling and cruising in formation were to be even more serious problems for the Naval Review Fleet. A new signal system, which had been worked out in the North Atlantic Squadron, had been approved by the Bureau of Navigation and was to be used by the Naval Review Fleet to supplement the standard Navy code.[94] Difficulties in communication were compounded by the fact that the various different types of vessels required different rudder angles to describe the agreed-upon standard turning circle of 2,730 feet. Gherardi had two buoys placed in Lynhaven Roads exactly 2,840 feet apart and ordered his ships to describe a perfect half circle between them (an extra 110 feet allowed so that the ships would not run the buoys over). They were to note the rudder angle required to perform this maneuver, which would then be known as that ship's "standard full rudder." The fact that Gherardi had to devote time prior to the review to work on details such as this shows a Navy that was still unaccustomed, even in 1893, to operating in formation.

On 4 April Gherardi ordered every ship currently at Hampton to sea for four days of exercises under the overall command of Rear Admiral Walker. While at sea Walker was ordered to "exercise [his] Squadron as per paragraphs eight and nine, of Section five, 'Programme for Naval Review.'" Specifically, they were to drill at getting under way together, keeping position when steaming in column, changing direction two and four points, and anchoring together.[95] Squadron commanders were also to ascertain the number of revolutions necessary to maintain a set speed of eight knots and the standard full helm, using the method described above. Competence at each of these tasks by the officers and crews of each ship would be important if all twelve ships were to maneuver together without mishap during the naval review.[96]

Gherardi's orders to Walker on this occasion make very interesting reading, more for what was not said than what was printed in the order. They were detailed, explicit, and professionally correct to the letter. Walker is specifically instructed to "signal for permission to get underway" and, later, to "make the necessary anchoring signals, but signal for permission to anchor." These were both signals that would have been professionally expected without a prompt, but Gherardi was taking no chances and was explicit about putting Walker in his place. Rear Admiral Benham was sent out to do essentially the same thing with his First Squadron on 11 April. Gherardi's orders to Benham also spell out various things the commander in chief wanted done as far as getting under way and anchoring, but the phrase "signal for permission" is never used. Reading the two sets of orders side by side (Walker and Benham probably did not see each other's orders) makes it clear that Gherardi wanted no repetition of the ugly business with Walker in Haiti in 1891, hence the specific language in Walker's version.

Both squadrons were sent out on 11 April, drilling independently under the command of their respective squadron commanders. Upon their return to Hampton Roads on 14 April, they anchored in their rendezvous formation, each ship riding to a single anchor, two cables (about 480 yards) apart. The "United States Fleet," as Gherardi had taken to calling it in his correspondence, was ready for the review. It was joined in Hampton Roads by the naval representatives of the other participating nations. On 24 April, it was time for the international fleet to begin its journey to New York. The United States Fleet got under way first, weighing anchor at 9:00 a.m. The twelve ships were reviewed by the secretary of the navy as they passed *Dolphin,* then they passed between the visiting squadrons, which were still at anchor, as they proceeded to sea. There is no doubt that this was done with a purpose. The Navy that twenty years earlier had been mortified that any foreigners would see the motley collection of wooden cruising vessels at Key West now not only had four credentialed correspondents embarked but was pointedly passing in review in front of an assembly of modern warships from major naval countries.[97] In this display the United States was proclaiming its new international status.[98] The assembled U.S. warships compared favorably in appearance to the various protected and unprotected cruisers of other countries. This was a relatively new state of affairs. Even five years earlier the warships of the North Atlantic Squadron had consisted entirely of old Navy wooden cruising vessels. The *Army and Navy Journal* recognized this. "What an exhibition we should have made of ourselves had Columbus landed a few years earlier so as to bring the four hundredth anniversary of his landing within the eighties," noted an editorial.[99]

The fleet steamed in two columns up the East Coast, the American, Dutch, and German ships on the port side and British, Russian, French, Italian and Brazilian vessels on the starboard. The two columns were six hundred yards apart, with an interval of three hundred yards between ships in column.[100] While this made an impressive sight during the day, it was not a safe formation for night-time transit, so as evening approached, *Philadelphia* signaled for the fleet to form columns of squadrons, each led by its own flagship, for the night. The night passed uneventfully, and the review columns were formed again in the morning.

As the fleet arrived in the waters of New York, they were met first by the Argentine flagship *9th of July* and later by the Russian flagship *Dimitri Donskoi,* which were already in New York. Also waiting for Rear Admiral Gherardi's arrival was the Patrol Division of the Naval Review Fleet, commanded by Capt. Frederick Rodgers. Captain Rodgers had set up a headquarters at the Army Building on Whitehall Street in New York City and commanded the movements of the torpedo boat *Stiletto* and every Navy tug that could be spared on the East Coast, as well as Revenue Marine (the forerunner of today's Coast Guard) steamers, Light House steamers, Naval Reserve and police tugs, and other service craft.[101] After a luncheon for dignitaries and the flag and commanding officers of the fleet on board *Dolphin,* newly elected President Grover Cleveland boarded a launch and was landed near Forty-eighth Street. At this point the Patrol Division, which had been keeping civilian sightseeing boats away from the fleet, removed their cordon, and the civilian craft flocked around the warships. That evening the spotlights of each ship lit up the harbor in a fascinating display of electrical power and pyrotechnics. This was followed on 28 April by the much-anticipated and unprecedented parade through New York of sailors from the international fleet.[102]

Rear Admiral Benham Takes Over

As the celebration drew to a close, the Navy Department had decisions to make about the future assignments, not only of all the ships concentrated in New York but also of the various flag officers involved. The disposition of the United States Fleet and the flag officer assignments after the naval review had been a subject of open speculation throughout the spring. As the senior officer in the Navy, having been at sea for over four years with the North Atlantic Squadron, the Squadron for Special Service, and the Naval Review Fleet, it would have made sense that the successful completion of the International Naval Review would have signaled the end of Rear Admiral Gherardi's career. It was reported, however, that he wished to remain on active duty until his statutory retirement for age in 1894, and that the command he wanted was his old North Atlantic Squadron. This, of course, would place him directly in conflict with his old rival, Rear Adm. John Grimes Walker.[103] Secretary of the Navy Hilary A. Herbert, with barely three months on the job, had a political minefield to negotiate, and he settled the question by giving none of the flag officers exactly what they had requested. Rear Admiral Gherardi became the commandant of the New York Navy Yard, relieving Captain Erben, who was scheduled to take *Chicago* to European waters to become commander in chief, European Station. Rear Admiral Walker was sent on three months' leave to await orders, and Rear Admiral Benham, who had been CinC of the South Atlantic Station, was named the new CinC of the North Atlantic Station.[104]

The disposition of the ships elicited almost as much interest as the flag officer assignments. The Naval Review Fleet represented the best and newest ships the New Steel Navy had to offer, and where the Navy Department decided to station them would speak volumes about the Cleveland administration's priorities.[105] The decisions eventually arrived at created a situation that prompted the *New York Times* to remark, "At no time within recent years has the United States Government been so well represented in foreign waters by an armed naval force, nor so poorly provided for in ships at home."[106]

Immediately after the naval review, five of the twelve ships constituting the United States Fleet were sent to navy yards to be fitted out for deployment. There was some concern that it might appear impolite for so many of the U.S. warships to depart the anchorage before the visitors had left, but Assistant Secretary of the Navy William McAdoo hastened to explain that the seven ships left could handle all of the entertaining that would be expected for the foreign visitors.[107] Eventually *Philadelphia* was sent to Honolulu, Hawaii, where naval troops from *Boston* had just figured decisively in the overthrow of the Hawaiian monarchy. *Yorktown* and *Charleston* were sent to the Pacific Squadron as well. *Newark* went back to the South American Squadron, and *Chicago,* as mentioned previously, went to Europe as the flagship of the European Station, taking *Bennington* with her. *Concord* went to the Asiatic Squadron, and *Bancroft* was sent to the Naval Academy to be used as a training ship for the naval cadets. *Atlanta* was put out of commission.

By the end of 1893 all of these moves nominally left four ships for the Home Squadron: The flagship, *San Francisco; Kearsarge,* which was one of the last of the wooden cruisers; the double-ended monitor *Miantonomoh;* and *Vesuvius,* the experimental dynamite cruiser whose 15-inch pneumatic guns were of questionable value in actual combat. In spite of the popularity of Mahan's new theories of seapower, the decision had been made to scatter the New Steel Navy throughout the world, in support of U.S. trade interests. "It is intended," reported the *New York Times,* "to keep ships of war in the waters of countries where there is a chance of increasing American trade. . . . The idea is based on the theory that American interests will be respected when an American cruiser is nearby."[108] In many ways, it was an unsatisfying end to the magnificent "coming-out" party of the United States Fleet.

Conclusions

On 31 May 1893, Rear Admiral Gherardi hauled down his flag as commander in chief of the Naval Review Fleet and moved ashore to take over as commandant of the New York Navy Yard and Station. Rear Adm. A. E. K Benham became the commander in chief the North Atlantic Squadron, on board his flagship *San Francisco.* The rest of the fall of 1893 remained relatively quiet. *San Francisco* traveled to Newport, Rhode Island, in August for the opening of that year's Naval War College class and was joined by *Miantonomoh* and *Vesuvius,* the former taking the Rhode Island naval militia on board for their yearly training.

Less than two months after the naval review, an event occurred that shocked the naval community. On 22 June 1893 HMS *Victoria,* the flagship of the British

Mediterranean Fleet, collided with HMS *Camperdown* during a tactical maneuver. She sank in minutes, taking most of her crew with her. A Royal Navy inquiry placed the blame for the disaster squarely on Vice Adm. Sir George Tryon, who had ordered a complicated maneuver without enough space between ships to carry it out. U.S. naval officers used the occasion to point out: "The necessity of constant practice . . . to prevent just such calamities in time of war and to familiarize our officers and men with the exact turning radius of each ship."[109]

In spite of this lesson, the squadron after the naval review was not much of a squadron. By October 1893, it consisted of Benham's flagship, the protected cruiser *San Francisco,* which represented the only operational modern warship of the squadron; the double-turreted monitor *Miantonomoh,* which never strayed far from her New York home port; and the venerable wooden cruiser *Kearsarge,* a relic suitable only to show the flag in Caribbean ports. The Navy Department was unsure what to make of the yacht-like appearance and unconventional pneumatic guns of the experimental *Vesuvius,* and consequently she spent much of her time doing utility work, finding and destroying hazards to navigation along the East Coast. The newly commissioned *Machias* was in Norfolk fitting out for an eventual cruise to the West Indies.[110] No two of these ships were alike, had the same functions, or had been designed with any thought to their being used together. When Benham left New York for the yearly cruise to the West Indies, his flagship went by itself. In fact, in his monthly report to the Navy Department, Benham confessed that the report was only accurate up to 18 October. "Since the latter date I have not been informed of their distribution or employment."[111] His December report was worse. "I have no knowledge of the distribution or employment of the other ships of the squadron during the month. The evidence is clear that by 1893, the North Atlantic Squadron was a fighting unit in name only.

Admiral Benham's plans for a peaceful winter cruise from port to port in the Caribbean were dashed when revolution broke out in Brazil. *San Francisco* sailed on 21 December 1893 for Rio de Janeiro. Captain Philip of the newly commissioned *New York* received a telegram on 26 December ordering her to sea immediately from New York. These two ships were eventually joined in Rio de Janeiro by *Charleston, Newark, Detroit,* and *Yantic.* Upon his arrival Rear Admiral Benham became the commander in chief of the South Atlantic Station while the incumbent, Rear Adm. O. F. Stanton, proceeded north to act as a caretaker for the eviscerated North Atlantic Squadron until his retirement in August. Training for the various naval militias took place during the summer, but otherwise no tactical exercises were conducted during 1894. For much of this time, most of the Navy's modern assets were with Benham in the South Atlantic.

The years 1892–94 had seen many steps forward in the process of developing a concept of multiship operations for the U.S. Navy. The Squadron for Special

Service and the Naval Review Fleet gave an entire generation of officers of the New Steel Navy vital experience in operating warships in formation. The Navy Department worked through complications with structural components of a fleet, such as a clear chain of command and a reliable logistics network. The international community, as well as the American public, gained a new image of an ascendant United States, viewed through the pageantry of the International Naval Review. The process, however, was not complete.

The development of processes critical to the formation of a multiship fighting capability, namely, tactics and doctrine, stagnated during this time. The Navy Department's attention was drawn instead to the political and diplomatic requirements associated with the Squadron for Special Service's many official visits during its trip around Cape Horn and the International Naval Review. These shortcomings, however, were about to be addressed.

With enough vessels of the New Steel Navy now in commission to cover the various stations, Secretary of the Navy Herbert announced a new squadron policy in 1894. Noting that "heretofore an insufficiency of numbers has, in cases of sudden emergency abroad, necessitated sending vessels from one station to another," he declared that from then on, he would "keep a number of cruising vessels sufficient for the ordinary needs of naval policing on each of the six stations":

> This policy will allow frequent fleet and squadron evolutions, which are absolutely necessary for the instruction of officers and men. To the North Atlantic, or home squadron, a sufficient number of vessels will be assigned to permit of a number being employed in practical exercises connected with the course of instruction at the Naval War College. Vessels fitting out on this coast will generally remain attached to the North Atlantic Squadron for the first six months of their cruise, for purposes of instruction and to enable officers and men to familiarize themselves with their ships. The home squadron will thus become the feeder for all the other squadrons.[112]

This was Stephen B. Luce's vision. He had been unable, during his uniformed career, to see it through to completion, but the stage was now set for his theories to become reality. The process of constructing an identity as a fighting force could only occur through rigorous exercise at sea. The conditions were now favorable for such an operational program, dedicated to training the North Atlantic Squadron to operate as a combat unit.

5

★

Luce's Vision Realized: The North Atlantic Squadron Solidifies a New Identity, 1895–1897

The advances in multiship operations under Rear Admiral Gherardi greatly increased the logistical and operational ability of the Navy to field a formation of warships, but Gherardi's work had been largely concerned with appearance, not tactical skills. It remained for the North Atlantic Squadron to develop a greater measure of the substance of a multiship fighting capability. By 1895 the Caribbean, the North Atlantic Squadron's traditional area of operations, was troubled by the renewal of armed resistance against the Spanish colonial authorities in Cuba. Elsewhere, both Great Britain and Germany threatened the Monroe Doctrine with unilateral actions against Caribbean governments. As the warships that had been authorized in the late 1880s began to arrive in the squadron, the commander in chief, for the first time, possessed the ability to concentrate enough vessels to regularly hold productive tactical exercises.

Squadron Cruise to the West Indies, 1895

The warships of the North Atlantic Squadron had cruised the West Indies regularly since the establishment of the station in 1865. Political and military requirements in the West Indies had often created roadblocks to effective squadron training, as the presence of one or more vessels was often urgently requested by the U.S. consulate whenever unrest broke out. The Navy Department, under pressure from the Department of State, typically had stripped warships from the Home Squadron to fulfill these requests. The winter of 1895, however, would be different. For the first time in the thirty years of the station's existence, in the absence of a crisis requiring separate deployments, the warships of the North Atlantic Squadron would visit the ports of the West Indies—"showing the flag" as a squadron, not as individual cruising vessels.[1]

From his flagship *New York,* berthed at the navy yard in her namesake city, Rear Adm. R. W. Meade made preparations for the cruise. His squadron consisted of the flagship, the new protected cruisers *Columbia* and *Montgomery, Atlanta,* and the experimental dynamite cruiser *Vesuvius. Columbia* was the namesake of a new class of lightly protected but speedy cruisers designed specifically for commerce raiding. She and her sister ship, *Minneapolis,* both built by the Philadelphia firm Cramp and Sons, were fast for the time at just under twenty-three knots, but the machinery necessary to drive the unique triple-screw engines made for a cramped and uncomfortable ship when at sea for extended periods. They were unpopular with sailors and officers alike, including Rear Admiral Meade.[2] *Cincinnati* and *Raleigh* were smaller protected cruisers. Authorized in 1888 and commissioned in 1894, they displaced 3,213 tons (about half the size of *Columbia*) and had a top speed of nineteen knots. *Montgomery,* launched in June 1894, was technically listed as an "unprotected cruiser," but she was often referred to as a "peace cruiser," the implication being that she was suited for peacetime patrolling duties but not engagement in combat with major warships. With a top speed of nineteen knots and a mixed armament of 6-inch and 5-inch guns, *Montgomery* was essentially a slightly larger version of a gunboat. The squadron was rounded out by *Atlanta,* one of the original ABCD ships, and *Vesuvius.* The latter was technically referred to as a "cruiser," but at 929 tons she was really more of an unarmored experimental gun platform than a warship. Taken together, the seven warships that made up the North Atlantic Squadron represented a national naval strategy that emphasized the traditional functions of cruising and attacks on commerce. This was not, however, how the department intended to utilize them in 1895.

Secretary of the Navy Herbert's and Rear Admiral Meade's objectives for this deployment were to be unlike what the annual secretary of the navy report had come to call the "usual winter's cruise." As Meade told a newspaper reporter, "Everything regarding the capabilities and weaknesses of the new navy is as yet experimental . . . a fleet hastily assembled and untried in fleet tactics meeting one that has been well drilled is a fleet destroyed."[3]

In order to have a squadron that could perform tactical maneuvers together, it was first necessary to get rid of warships that would be unable to keep up with the most modern vessels.[4] A request from the city of New Orleans to have a representative of the New Steel Navy at their Mardi Gras celebration provided the perfect excuse to detach *Atlanta,* which was already at Hampton Roads.[5] Now nine years old, her horizontal compound engines were a generation behind the vertical triple expansion engines of the newer cruisers, and her top speed of fifteen knots was four to eight knots slower than that of the other ships in the squadron. In addition, Rear Admiral Meade apparently did not like her commanding officer.[6] *Vesuvius* was detached from the North Atlantic Squadron on 2 January 1895 and detailed by the Navy Department to search for and destroy partially submerged wrecks and other hazards to navigation along the East Coast.

Rear Admiral Meade departed Hampton Roads as soon as he was able to collect *Cincinnati* and *Raleigh.* The three ships left on 30 January 1895.[7] *Minneapolis* and *Columbia* were to follow when their yard work was completed. The squadron made for St. Thomas, Dutch West Indies, stopping at Samana Bay in the Dominican Republic along the way to check for the rumored presence of the French fleet. Meade found much to satisfy him on his squadron's first underway evolution together. On the way to St. Thomas, the ships exercised at maneuvers and held signal drills.[8] Meade had observations to make about both. On the whole he was pleased with the performance of his flagship, *New York,* although he felt that she had too much woodwork and other flammable materials that needed to be removed after her current commission was up.[9] He was, however, less impressed with the protected cruisers. These he criticized as difficult to steer handily and inefficient in the burning of coal. His problems as a commander in chief also included the inability to send signals efficiently, both night and day. "On this trip I made twelve pages of fools-cap [scratch paper]," lamented Meade, "of signals important at this day that are absolutely ignored in the General Signal Book, each of which should be made in one hoist of flags."[10] The more time his warships spent in company with the flagship, the more the inadequacy of the Navy's signaling system was exposed.

From St. Thomas, the three ships went to St. Croix then St. Lucia before stopping at St. Pierre, Martinique, then St. Lucia, and finally Bridgetown, Barbados. Here, the squadron paused to celebrate Washington's birthday. The holiday provided another opportunity for the squadron to grow together as a team. A series of boat races was organized, including a printed program outlining the rules and the different classes of boats for each race. There was nothing new about marking holidays with boat races between ships that happened to be in port together, but this particular event, with its printed program advertising the races of the "United States Squadron," provides another small piece of evidence pointing to the solidification of a permanent identity for the North Atlantic Squadron.[11]

The squadron left for Port-au-Spain, Trinidad, on 28 February, arriving on 1 March. Although Rear Admiral Meade was irritated that the local authorities in Barbados had not acknowledged Washington's birthday, otherwise the visit there had been a success. The U.S. consul noted that the local officials had arranged a ball for the officers of the squadron—the first time he was aware of that "any public entertainment has been given in honor of the visit of foreign warships." Meade sent a proper letter of thanks to the committee in charge of the dance, which was reprinted in the local paper. If he was unhappy with the authorities, he at least kept it to himself while he was there.

In his 1 March situation report, Meade praised the idea of having the three ships cruise together, boasting, "I am informed by officers who served in the White Squadron that the NEW YORK, RALEIGH, and CINCINNATI keep better line and column, and maneuver better than the White Squadron did after six months practice."[12] He went on to properly give some of the credit for this to his ships' twin-screw designs, but his delight in the squadron's prowess was unmistakable.[13]

He was also proud of their signaling ability, which he exercised daily, but was of the opinion that the current system of signaling would have to be improved upon if a number of vessels were going to operate together and maneuver rapidly.[14] Meade's chief of staff, *New York's* commanding officer, Capt. Robley Evans, noted that "Admiral Meade . . . gave us admirable and systematic drill. Modern methods and appliances were used in a modern way—torpedoes were run under service conditions and searchlights used to their utmost capacity as a means of communicating."[15]

The squadron's stay in Port-au-Spain took a dramatic turn on 4 March when a large fire was spotted downtown. Rear Admiral Meade immediately gave orders for all three of his ships to prepare fire parties to go ashore and dispatched an officer to the U.S. consulate with an offer of assistance. There was some delay in sending the boats ashore as Meade was concerned about landing troops on foreign territory without a formal request for assistance from the local authorities, but as the fire grew in intensity, Chief of Staff Evans took the initiative to order about 225 men from all three ships to land in the city and provide assistance.[16] After about four hours ashore battling the fire, the men were able to turn the tide and the rest of the city was saved. The next day, a letter from the governor arrived on board *New York,* profusely thanking the men of the North Atlantic Squadron for their timely assistance. Meade promptly had the letter copied into a circular order and ordered it read to all hands at quarters. It was a great example of squadron teamwork, and what Captain Evans referred to as "the most important work of Admiral Meade's squadron during the West India cruise."[17]

While the city of Port-au-Spain was still smoking, *Minneapolis* arrived from the United States, fresh from having her torpedo outfit installed at Newport. The Navy Department detached a member of the squadron on 6 March, when it cabled orders for Meade to send *Raleigh* to Colón to reinforce *Atlanta.* The Nicaraguan rebels had attacked Bocas del Toro, about one hundred miles north of Colón. *Atlanta* was ordered to Bocas del Toro to protect the considerable U.S. commercial interests there, while *Raleigh* was ordered to secure the Atlantic end of the Panama Rail Road at Colón.[18] Like other commanders in chief before him, Meade took it personally that one of "his" ships was being sent on detached duty and in his usual outspoken manner expressed his views to the department: "I regret very much to lose her at this time. . . . The sister ships were beginning to show the effects of drill together in squadron. . . . I trust the reports received by the Department from Colón were accurate enough to justify the orders."[19]

Coal continued to be an issue for the squadron, demonstrating once again evidence that, logistically, the United States did not have the resources in place to support a large concentration of warships. On 15 March, Captain Cromwell of *Atlanta* sent *Raleigh* from Colón to Cartagena to take on coal. It had to be purchased from a civilian company, the Cartagena Terminal and Improvement

Company. Captain Miller reported to the Navy Department that the company had about one thousand tons of coal "to spare" and that he could purchase it at $5.45 per ton.[20] The department's reliance on private suppliers and the good graces of local businesses to fuel their warships was a major hindrance to having more than one ship in any port at a given time.

While *Raleigh* was busy refueling, the rest of the squadron departed La Guayra, Venezuela, on 18 March. *Cincinnati* was detached and sent to Curacao for mail and dispatches, rendezvousing with the rest of the squadron off Little Curacao Island Light on the morning of 19 March, and the three ships continued on to the Dominican Republic. Along the way, Rear Admiral Meade exercised the squadron in distant signaling with searchlights and Very signals. The squadron arrived at Santo Domingo City on 20 March 1895.

On 22 March, Admiral Meade was presented to the president of the Dominican Republic by the American consul. President Heureaux was concerned about the French squadron patrolling the Caribbean and more than a little grateful for the presence of U.S. warships and for the good offices of the U.S. government in the ongoing trouble between France and the Dominican Republic.[21] Meade replied that he and his squadron were "in these waters for the protection of American interests . . . and that I could assure him it was my intention to fully protect the interests of all citizens of the United States." Whether or not Meade offered President Heureaux direct protection against the French fleet, his presence was still reassuring.

The squadron departed for Kingston, Jamaica, bypassing a planned port visit at Port-au-Prince, Haiti, much to the chagrin of the U.S. minister there, who complained to the secretary of state about all the disappointed U.S. businessmen in Haiti who would have benefited from the visit of the North Atlantic Squadron. The squadron arrived at Kingston on 24 March, having exercised at fleet maneuvers and signaling on the way. After exchanging the usual visits and salutes, the squadron was met on 27 March by *Columbia,* carrying Assistant Secretary of the Navy McAdoo. Much of the squadron's time over the next week was spent coaling. The Navy Department had hired colliers to bring U.S. coal to the squadron at Jamaica. This was an improvement over trying to get coal either from local firms at inflated prices or from U.S.-owned firms who might have some coal to spare. It was still a headache, however, as the department had sent too much coal.[22]

The squadron left Jamaica on 8 April, proceeding to Port-au-Prince. Once again, Meade had issues with the amount of time the local authorities took to grant the squadron clearance to enter port. The delay of three hours and fifteen minutes earned the U.S. minister resident a letter from Meade asking for an explanation. The small hints at Meade's temper in his official correspondence help the reader understand the Navy Department's reluctance to allow him much interaction with foreign officials. In Port-au-Prince, Meade got word that the French flagship *Duquesne* was possibly headed toward Santo Domingo City to investigate the murder of a French citizen at Samana Bay and press the French government's demands for payment of debt. Meade ordered Captain Wadleigh and the *Minneapolis* to Santo Domingo City to keep watch on the French admiral's actions and report back. Meanwhile, the department seemed reluctant to allow Meade to become involved—even as an observer. Meade was cabled, countermanding his orders. He was told to leave the French alone and proceed to Colón, Colombia.[23]

Meade's temper was about to get him in worse trouble. Before he left Port-au-Prince, he answered Secretary Herbert's letter of 9 March, which had reprimanded him for some earlier remarks he had made about the British colonial officials at Barbados failing to observe Washington's birthday. Meade's reply to the secretary's disapproving letter was a single sentence: "The information contained therein is very interesting." He seemed unable to acknowledge an error in judgment and move on, and his snide reply of 10 April infuriated Secretary Herbert. Herbert, in turn, dictated a letter to Meade dated 19 April demanding an explanation in writing of what Meade meant by the remark "the information contained therein is very interesting."[24] Meade's reply of 24 April from Key West was no better than his first answer: "If I had stated that I found the information furnished by the Department not interesting, the Department might, with justice, have taken me to task."[25]

In the meantime, the squadron left Port-au-Prince for Colón on 11 April, arriving on the fourteenth. Meade again exercised his squadron at fleet maneuvers during the run from Haiti to Colón. The correspondence does not specify exactly what maneuvers the warships were performing, but according to Capt. Robley Evans, *New York*'s CO at the time, they were realistic and worthwhile.[26]

At Colón the squadron met *Atlanta* and *Raleigh*. Captain Cromwell reported all quiet since his arrival. Meade not only thought that things were quiet but also questioned just how much business interest U.S. citizens even had in Colón. He pointed out that most of the property of the Panama Rail Road was owned by the French and went on to say that the guarantee of the integrity of the transit across the isthmus mainly benefited the French: "These men and their employees are the people who constantly raise the cry that because there is a revolt . . . 800 miles off and up the country, that an American ship of war must be kept at Colón to protect American interests." He went on to say that "if the Navy Department could only realize what an extraordinary amount of humbug and self interest of the foreigners enters into this business, it might save much money for work more useful." Meade was referring to tactical formation work. He was still annoyed that *Raleigh* had been ordered away from the squadron while he was trying to exercise his warships back in March.[27] Meade saw the time spent on what he considered to be a "humbug" as a distraction from his ability to exercise his squadron in tactical work.[28]

Minneapolis was sent out on 20 April for her final speed trial, Captain Evans of *New York* again acting as the president of a board appointed by Meade to oversee the trials. That same day, *New York* and *Columbia* sailed for Key West. *Minneapolis* headed for Kingston, Jamaica, to take the rest of the coal from the schooner the Navy Department had sent there for the squadron. *Atlanta* and *Raleigh* were ordered to Key West as well, although those two were ordered to proceed separately, as *Atlanta*'s slow cruising speed (about eight knots) would make it uneconomical for her to sail with the flagship.[29]

The deployment of the North Atlantic Squadron to the West Indies ended with the arrival of the flagship *New York* off Key West on 24 April 1895. It had been a historic cruise. While individual ships had been ordered off on Navy Department tasking, on the whole the squadron operated as a unit. Meade had emphasized training, formation work whenever possible, and night and day signaling. In fact, on the run in to Key West, Meade reported on exchanging search light signals with *Cincinnati* at ranges up to thirty miles.[30]

Rear Admiral Meade Retires

Rear Admiral Meade arrived back in New York on board his flagship on 28 April. He reported his arrival and went on to express his "great disappointment at the virtual breaking up of this squadron just at a time when it was getting into promising condition and especially do I regret that I could not have had the drills in Florida Bay that I projected three months since."[31] The Navy Department had other concerns, however. In Germany, Kaiser Wilhelm II was preparing to celebrate the opening of the Kiel Canal. This strategically important waterway linked the North Sea with the Baltic Sea, allowing ships to save about 250 miles by not having to transit around Denmark when moving to or from the Baltic Sea. To represent the United States at the naval celebration, the Navy Department decided to send *New York* and *Columbia*.

The loss of his flagship frustrated Meade even more than having other warships under his command ordered away for Navy Department errands. He sent a petulant letter, complaining among other things that his band instruments and typewriter were going to be damaged while moving from ship to ship and requesting, if he absolutely had to move, to be allowed to shift his flag to *Cincinnati*.[32] Meade thought particularly highly of Captain Glass, *Cincinnati*'s CO, and it was probably not a surprise at the Navy Department that he wanted to shift his flag to her, but she had been ordered by the Bureau of Construction to Norfolk for repairs and refitting. Because work at the shipyards was slow, the bureau was trying to distribute the Navy's needed repair work evenly, and the New York Navy Yard already had several ships of the squadron in hand.

Capt. William B. Cushing (1842–1874) epitomizes the change that is the argument of this book. Within two months he went from autonomously carrying out national policy in a foreign port (using the threat of his ship's cannons and armed sailors) to steaming in a formation of ships responding to the direction of a flag officer. *U.S. Naval Institute Photo Archive*

This photo of the USS *Wyoming* (commissioned in 1859, scrapped in 1892) was probably taken in the late 1880s, when she was used for training midshipmen at Annapolis. Imagine attempting to conduct tactical exercises under steam with a fleet of vessels like this one. *U.S. Naval Institute Photo Archive*

Rear Adm. Augustus L. Case (1812–1893) was a former chief of the Bureau of Ordnance and the commander in chief of the European Station when he was called upon to lead the U.S. Navy response to the *Virginius* affair and the subsequent exercises off Key West in 1874. *U.S. Naval Institute Photo Archive*

The first warship of the New Steel Navy, the USS *Atlanta* (commissioned in 1886, sold in 1912), is pictured here in Boston, circa 1889, while part of Rear Adm. Walker's Squadron of Evolution. *U.S. Naval Institute Photo Archive*

Commo. Foxhall A. Parker Jr. (1821–1879) was the U.S. Navy's post–Civil War expert on tactics and formation steaming. He undoubtedly would have been more prominent in the early years of the New Steel Navy had he not died suddenly in 1879 while serving as superintendent of the United States Naval Academy. *U.S. Naval Institute Photo Archive*

Rear Adm. James E. Jouett (1826–1902) was commander in chief of the North Atlantic Squadron from 1884 to 1886. Jouette is most well known for leading the U.S. intervention in Panama in 1885. *U.S. Naval Institute Photo Archive*

Rear Adm. Stephen B. Luce (1827–1917) is best known as the "Father of the Naval War College." His constant desire to synthesize classroom learning with rigorous operational exercise is often overlooked. *U.S. Naval Institute Photo Archive*

Rear Adm. John G. Walker (1835–1907) was opinionated, egotistical, and very well connected. In many ways Walker was the perfect flag officer to act as the operational trailblazer for the New Steel Navy in its infancy. *U.S. Naval Institute Photo Archive*

Rear Adm. Bancroft Gherardi (1832–1903) was often torn between his "old-fashioned" duties as a warrior-diplomat and the modern demands of command of a concentrated body of warships. *U.S. Naval Institute Photo Archive*

The Squadron of Evolution, circa 1889. This photograph raises the question: Why did the Navy construct *cruising* vessels in the 1880s, then assign them together as a *squadron*? *U.S. Naval Institute Photo Archive*

The USS *Kearsarge* (commissioned in 1862, struck in 1894), a popular Civil War veteran, saw long service in the North Atlantic Squadron. As the squadron modernized and began to do more formation work, she became the preferred vessel for detached operations in the Caribbean. *U.S. Naval Institute Photo Archive*

The USS *Philadelphia* (commissioned in 1890, struck in 1926) was one of the second generation of protected cruisers, constructed after the so-called ABCD ships. Upon her arrival in the North Atlantic Squadron, Rear Adm. Gherardi made her his flagship. She continued service in this capacity for several years. *U.S. Naval Institute Photo Archive*

Rear Adm. Richard W. Meade III (1837–1897), the nephew of the more famous Army general, was commander in chief of the North Atlantic Squadron, 1894–1895. His temper and confrontational manner made him ill-suited for the requirements of squadron command in the New Steel Navy and led to his early retirement. *U.S. Naval Institute Photo Archive*

The 1893 Naval Review in New York, a true United States "fleet" under Rear Adm. Bancroft Gherardi, comprised two subordinate squadrons. The review was the largest concentration of U.S. Navy ships since the 1874 Key West exercises. Contrast the pride with which the U.S. Navy displayed its warships to the world in 1893 with the embarrassment many officers felt during and after the 1874 exercises. *U.S. Naval Institute Photo Archive*

The USS *New York* (commissioned in 1893, struck in 1938) was the first of the armored cruisers. She served as the North Atlantic Squadron flagship for several years, including flying Rear Adm. Sampson's flag at the Battle of Santiago in 1898. This photo was taken immediately after the battle. *U.S. Naval Institute Photo Archive*

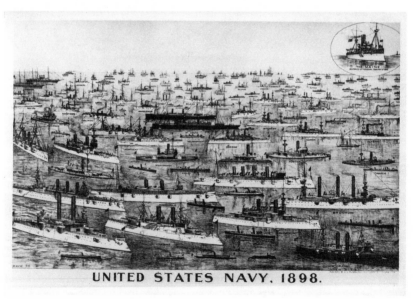

The U.S. Navy in 1898. All of the modern warships of the North Atlantic Squadron can be seen in this contemporary lithograph that captures the transition to the New Steel Navy. Note some of the older ships in the background. *U.S. Naval Institute Photo Archive*

None of this mattered to Meade, who essentially demanded *Cincinnati*. Amazingly, the department relented, going so far as to have a pilot boat intercept *Cincinnati* off Hampton Roads and tell Captain Glass to steam directly to New York.[33] It was, however, too late. While those orders were being given, but before they could be carried out, Meade asked to be relieved of command. In nine and a half months of command, Meade had never been able to find his stride. From his exchange of disrespectful letters with the secretary of the navy, to his constant complaints over the employment of his warships, he was unable to function effectively as a commander in chief in the new Navy.[34]

At first the papers reported that he was retiring for his health and desire to take a European trip.[35] But having resigned his command, Meade could not resist public declarations. When a reporter for the *New York Tribune* cornered him, he vented his frustration: "My ideas are not in accordance with those of this Administration. I am just as much disgusted with it as the people at large in the country are, and I preferred to quit rather than continue my association with it. . . . I am an American and a Union man. Those are two things that this Administration cannot stand."[36] It was outright insubordination—not only to Secretary Herbert (a former Confederate) but also to President Cleveland himself. For a while, it appeared that Meade would face court-martial, but eventually, the president allowed him to retire quietly to New York.[37]

Rear Admiral Bunce and Squadron Exercises, 1895

Command of the squadron devolved to Rear Adm. Francis M. Bunce. Bunce was a native of Hartford, Connecticut. Appointed to the Naval Academy in 1852 at the age of sixteen, he had graduated in 1857. As a lieutenant during the Civil War, he served in a variety of positions that gave him experience both in small boat and disembarked operations, as well as the more traditionally squadron-oriented blockading duty. His Civil War service culminated in command of the monitor *Monadnock,* which he took around Cape Horn, steaming from Philadelphia to San Francisco, after the war ended. It was the first extended sea voyage ever made by a monitor. Prior to taking command of the North Atlantic Squadron, he commanded the Naval Training Station and the training ship *Richmond* at Newport, Rhode Island.

Within weeks of taking command, Bunce had proposed a training schedule for the squadron. The department had requested the squadron's presence at Portland, Maine, no later than 26 August 1895. Bunce proposed leaving New York on 1 August and sailing to Gardiner's Bay for target practice. The squadron would then proceed to Newport, Rhode Island, for exercises in support of the Naval War College class graduating that term. After a call at Bar Harbor, Maine, they were to arrive at Portland, Maine, by 25 August. Bunce was anxious to exercise his new command at fleet tactics and stipulated that fleet tactics at sea would be exercised en route between ports.[38]

Fleet tactics would share top billing in the summer and fall of 1895 with the exercise of the various naval militia units up and down the East Coast. Bunce detailed the monitor *Amphitrite* to carry out most of this tasking. This slow and unseaworthy vessel was not very useful for fleet tactical work, but her big guns made her an excellent training platform for the militia crews. *Amphitrite* was ordered to proceed to Brunswick, Georgia, where Captain Wise was to report to the adjutant general of Georgia and provide the Georgia naval militia with two days of drill. *Amphitrite* was then to proceed to Charleston, South Carolina, where the naval militia of South Carolina would be drilled, and finally to Southport, North Carolina, for five days of drill with the North Carolina naval militia.[39] While the rest of the squadron was preparing for the summer maneuvers along the East Coast, *Atlanta* was at Key West, Florida, enforcing President Cleveland's proclamation of strict U.S. neutrality in the ongoing revolution in Cuba.[40]

The squadron anchored together in Newport Harbor on 8 August 1895. They were met by Secretary Herbert, who had traveled to Newport in *Dolphin* for the commencement of the summer maneuvers.[41] Their presence in Newport marked the height of the Newport social season.[42] A gala reception awaited the officers of the squadron the Tuesday night after their arrival. On 13 August, Governor Lippitt of Rhode Island was hosted on board the *New York,* as were many other visitors.[43] It was not all festivities. While in Newport, the ships exercised boats, landing parties, and torpedoes and planned for the follow-on maneuvers that would take place at sites up and down the East Coast.[44] The idea was to "conduct a campaign such as would prevail during a war with foreign vessels endeavoring to capture cities along the Atlantic coast."[45]

After conferring with the Naval War College and completing their business in Newport, the squadron, now joined by *Raleigh,* got under way on 15 August for the resort town of Bar Harbor, Maine.[46] The secretary of the navy quietly took passage to Bar Harbor on *New York,* not raising his flag to be officially recognized on that vessel but simply observing. Along the way, the warships practiced every maneuver in the fleet drill book. Some of the more important ones were performed several times. In all, Rear Admiral Bunce reported that the "good results have begun to appear, already, in increasing uniformity in speed, in keeping distance, and in time of making turns and circles."[47] After exhaustive trial and error, the squadron concluded that their squadron tactical diameter should be 750 yards, based on the turning capability of *Minneapolis* at nine knots. In all it was a successful trip, not only from the point of view of the tactical work they were able to accomplish but also because of the social calls they were able to make in the various cities in the Northeast that they visited, as well as the time the officers of the squadron were able to spend together at social functions.

At Bar Harbor, Secretary Herbert disembarked from *New York* and went back on board *Dolphin*. The arrival of the North Atlantic Squadron had been eagerly awaited by the society luminaries who summered in Bar Harbor. Among the notables who waited to greet the squadron were John Jacob Astor, the retired secretary of state and sometime presidential candidate James G. Blaine, and the Rockefellers.[48] Again, the squadron officers were entertained at numerous society events, the highlight of which was the grand ball and reception given in honor of the secretary of the navy on 20 August.[49] The officers were able not only to socialize with the citizens of Bar Harbor but also to spend time with each other. The officers of the various ships had the opportunity to discuss the previous week's tactical exercises, exchanging information and observations about the drills. This socialization of the officer corps was a crucial component of the development of an identity as a fighting organization. Critics, however, felt that the North Atlantic Squadron was not doing enough. They complained that the warships of the South Atlantic and European stations had not been called to New York to assemble a U.S. fleet for massive exercises, simulating attacks on New York or Boston. The Navy Department largely shrugged off these complaints, in keeping with Secretary Herbert's stated 1894 policy of keeping squadrons together as much as possible on their respective stations. The old days of responding to every contingency with ad hoc concentrations of all available warships were being traded for the cohesion and professionalism to be found in the repeated, rigorous exercise of a single combat unit.[50]

The North Atlantic Squadron left Bar Harbor on 22 August, arriving at Portland, Maine, after another four days of intensive tactical drills. They were now routinely performing intricate maneuvers at nine knots during the daytime and six knots at night. While at Portland, the officers of the squadron had yet another opportunity to be entertained by the city notables.[51] After three days in Portland, the warships departed for Boston on 29 August.[52] By now newspaper articles were showing confusion over the new image and identity of the North Atlantic Squadron. Reporters were so used to squadron-sized operations being the result of an ad hoc concentration of warships for a specific contingency or purpose that they were frankly unsure of what to do with the fact that the summer maneuvers were simply the normal operations of the North Atlantic Squadron. Various reports in the *New York Times* began to refer to the "White Squadron," or the "Squadron of Evolution" as if there had to be some purpose for all these warships to be concentrated on the East Coast. Finally, the editors of the *Times* put an end to the confusion by promising in an editorial to stop referring to the North Atlantic Squadron as the White Squadron, as they pointed out that the rest of the ships in the Navy were white as well.[53]

The squadron departed Portland for Boston on 29 August, arriving the next day. The wharves were lined with hundreds of people anxious for a look at, or perhaps even a visit to, the white warships.[54] The landing exercises complete, the squadron weighed anchor from Fisher's Island Sound on 15 September and arrived at Tompkinsville, Staten Island, the next day, where they took on coal. Bunce intended for the squadron to spend about five days in Tompkinsville, replenishing supplies and coaling before departing for Virginia. They were joined there by *Columbia,* which had finished her repair period in the navy yard and was ready to rejoin the squadron. The arrival of the battleships *Texas* and *Maine* was also eagerly anticipated. These two second-class battleships had been authorized by Congress in 1886, before the armored cruiser *New York,* but construction delays had slowed their commissioning. *Texas* was the product of a design competition sponsored by the Navy Department and won by the Naval Construction and Armaments Company of Barrow-in-Furness, England. Construction took place at the Norfolk Navy Yard. At 6,315 tons displacement, she was smaller than most European battleships. Her main armament consisted of two 12-inch rifles in single turrets fore and aft, with a secondary battery of six 6-inch guns.

Maine was designed by the Navy Department and built at the New York Navy Yard. She was slightly larger than *Texas,* at 6,682 tons, with a main battery of four 10-inch guns mounted in two double turrets and a secondary battery of six 6-inch guns.[55] She was commissioned on 17 September, the day after the squadron's arrival back in New York, and her new commanding officer was Captain Crowninshield, who had previously served in the North Atlantic Squadron as the CO of *Kearsarge.* Crowninshield immediately reported by letter to Rear Admiral Bunce placing himself and the ship under his command at the admiral's orders.[56] It would take another four or five weeks for her to be ready to cruise. After provisioning, she had to take on ammunition then sail to Newport for her torpedo outfit. Bunce's plans for the summer exercises had envisioned three phases. The first, including the basic formation work and the landing exercises, was complete. With the addition of *Columbia, Texas,* and *Maine,* the second phase of maneuvers in the exercise grounds off Hampton Roads could start.

New York, Raleigh, Minneapolis, Columbia, and *Montgomery* got under way from Tompkinsville on 23 September for target practice and squadron drill, "repeating all former evolutions and performing such others in the Fleet Drill Book as are thought to be valuable for exercise or for use in action."[57] On each of the first two days out of New York, the five ships spent about two hours shifting between the basic formations of echelon, line abreast, and column.[58] The work got significantly more difficult as the squadron made its way south. On the third day, the ships worked for four hours, incorporating simultaneous turns and more intricate formation shifts. By the fourth day, they were breaking up into sections (two and three ships, respectively), forming into columns of sections, manipulating the distance between ships from close to open order, and turning the entire formation. The squadron worked for about four and a half hours on the fourth day and another four and a half on the fifth. With the exception of Gherardi's Naval Review Fleet, which was executing maneuvers strictly for appearance, formations this complicated with more than four ships had not been attempted since the Key West exercises in 1874. At that time, the formation could barely maintain a top speed of four knots. The ordered speed for most of Bunce's maneuvers was nine knots, which his warships had little trouble sustaining. After five days, the squadron came to anchor in Hampton Roads. The weather precluded the target practice that Bunce had hoped to hold in the Virginia Capes, but the work the squadron had done was impressive nonetheless. Bunce's official report of the maneuvers practiced ran to six single-space printed pages. "There is no evolution in the Fleet Drill Book," he wrote to the Bureau of Navigation, "that has not been tried by at least four of the ships now in company."[59]

Although the squadron did not hold target practice after arriving at Hampton Roads, Rear Admiral Bunce had issued the orders for the exercise that was to have taken place on 30 September. The orders give a glimpse of what stationary target practice entailed in 1895. The ships were to be anchored at one-thousand-yard intervals. Officers from each ship were to be sent to adjoining ships in order to observe the fall of shots from their ship's guns. Two broadsides were to have been fired from the main battery, and four from the secondary battery on the starboard side.[60] After the tide shifted and the ships swung around, the process would be repeated on the port side.[61]

The squadron remained at anchor off Hampton Roads for the next two weeks, with the exception of 18 October, when all the ships got under way and proceeded to Newport News Shipyard to be present at the launching of the new gunboats *Nashville* and *Wilmington*.[62] On 21 October the squadron weighed anchor and proceeded to the southern drill grounds, where they were finally able to hold the long-postponed target practice. Targets were anchored and the ships were detailed to observe each other. This evolution was carried out under way, the ships of the squadron proceeding at a base speed of nine knots, between 1,000 and 1,300 yards from the targets. Each ship made four runs by the targets, firing its main and secondary batteries separately.[63] Rear Admiral Bunce pronounced himself pleased with the results, noting that five targets were hit "under the conditions that would obtain in action." This may have been an optimistic comment, as the squadron's opponents in this "action" were anchored in place.[64]

After more tactical work, raising the formation speeds to twelve knots, the squadron ended its training exercises and broke up. *New York* returned to the New York Navy Yard, arriving on 9 November. *Columbia, Raleigh,* and *Montgomery* stayed at Hampton Roads, the latter two undergoing repairs at the Norfolk Navy Yard. *Minneapolis* was detached from the North Atlantic Squadron and ordered to the European Station on 21 November,[65] and *Amphitrite* was ordered to Key West to relieve *Cincinnati,* which had been there since August.[66]

Rear Admiral Bunce wasted little time preparing for his next endeavor. The year of training complete, he was now contemplating an operational deployment of his squadron to the West Indies. In comparison with squadron plans that had been submitted in past years, Bunce's submission of his proposed itinerary for the winter cruise did not list individual ships and the ports they might be sent to but simply assumed that the squadron would be operating as a unit. The Navy Department was apparently also anxious to exercise this capability, asking Bunce for a list of repairs to his ships that were absolutely necessary to make the deployment happen as soon as possible.[67] Coaling so many ships at the same time was still a problem for deployment, as Bunce assumed that he would have to break up the ships between St. Lucia and St. Thomas for refueling. The plan was to buy from local vendors on this cruise rather than send the coal down from the United States. Apparently the sending of colliers during the previous deployment had convinced local merchants to lower their prices acceptably. Bunce proposed being gone from about 21 December 1895 to 12 May 1896, touching at virtually every port in the Caribbean where there was a significant U.S. business interest.[68]

New York and *Columbia* got under way from New York for Hampton Roads on 16 December, arriving off Norfolk the next day.[69] There they were joined by *Montgomery* and *Raleigh*. *Maine* arrived on Christmas Day. On 29 December 1895, Rear Admiral Bunce reported the five warships ready for deployment.[70] The first day of January 1896 found the squadron at anchor off Hampton Roads. Celebrations began at 11:50 p.m. the night before, when the steam whistles of the ships joined in a cacophony leading up to midnight. Sailors and junior officers alike serenaded the flag quarters before turning in. The next morning after quarters, a "rope yarn" day was declared, and the sailors enjoyed rare time off on board ship. The junior officers traveled between ships in boats, wishing members of the other wardrooms a happy new year. It was evidence that the officers of the squadron had developed close ties over the previous six months of exhaustive maneuvers together.[71]

Plans, Contingencies, and Exercises, 1896

The squadron was not able to deploy as Rear Admiral Bunce had planned, however. Contingencies arose that superseded his carefully prepared peacetime itinerary. The renewed revolutionary unrest in Cuba made the Navy Department reluctant to dispatch warships to the Caribbean, lest the Spanish become uncomfortable.[72] Additionally, the long-running dispute between Great Britain and Venezuela over the border of British Guiana threatened to provoke British intervention, which was increasingly perceived as a threat to the Monroe Doctrine. In the summer of 1895, the Cleveland administration decided to act. Recent U.S. naval advances, including the North Atlantic Squadron's successful 1895 squadron deployment to the West Indies under Rear Admiral Meade, the anticipated summer and fall maneuvers, and the imminent additions of the battleships *Maine* and *Texas* almost certainly affected the decision. In July 1895, Secretary of State Olney sent a note to the British government reiterating the Monroe Doctrine, demanding that the British submit the boundary issue to arbitration, and containing the famous line that became emblematic of the age: "Today the United States is practically sovereign on this continent and its fiat is law upon the subjects to which it confines its interposition."[73] That sentence would have been unthinkable a decade earlier. The perceived effectiveness of the North Atlantic Squadron was the only reason it could be uttered in 1895. The British prime minister, Lord Salisbury, waited four months before replying to Olney's note, which gave newspapers plenty of time to discuss the merits of the Royal Navy versus the new U.S. Navy and critique U.S. shore defenses.[74]

Rear Admiral Bunce traveled to Washington, D.C., for consultations at the Navy Department.[75] It was decided to await the arrival of *Maine* and possibly *Texas* as well before the squadron would proceed. While the department was deciding on the best employment for the North Atlantic Squadron, another foreign relations crisis arose. The U.S. minister to Turkey made demands upon the Ottoman authorities in Constantinople for compensation for destruction of property belonging to U.S. missionaries working in Turkey. The strong words from the minister led to a round of newspaper speculation, again involving the North Atlantic Squadron. Stories ranged from the entire squadron being sent to the European Station (whose commander in chief, Rear Admiral Selfridge, was senior to Rear Admiral Bunce), to a few of Bunce's ships being sent to augment the European Squadron.[76] In the end nothing came of the trouble with Turkey, but it still served to delay the departure of the squadron on a cruise to the West Indies as planned.[77]

Maine arrived on 6 January 1896 and was immediately ordered to sea with one of the seasoned ships of the squadron, *Raleigh,* for two days of section drill. While the various political issues were playing out, it was up to Rear Admiral Bunce to keep his squadron, which was marking time off the coast of Norfolk, occupied. By February it became clear that the squadron was not going to be deployed immediately. One by one Bunce began to send his ships north for liberty. *Columbia* was the first to go, leaving on 4 February.[78] In April, with still no definite orders in sight, Bunce detached several of his ships to give the men some rest. *Columbia* went back to Staten Island, and *Montgomery* and *Cincinnati* went to the Norfolk Navy Yard. The flagship followed *Columbia* back to Tompkinsville a few days later, Bunce leaving explicit instructions for the senior officer present to govern the squadron in his absence.[79] Upon *New York's* return to Hampton Roads, *Newark* reported to the squadron, having been previously assigned to the South Atlantic Station.[80] Bunce sent her north to New York Navy Yard to be docked, have her bottom painted, and discharge sailors whose term of enlistment had expired while they were in South America.[81]

In lieu of deploying to the Caribbean, a series of exercises was planned for the late summer and approved by the department in June. Bunce was given permission to "make trials of such formations and movements as may seem . . . desirable."[82] On 1 August, Bunce reported *New York, Indiana, Cincinnati, Amphitrite, Newark,* and *Fern* ready for sea. *Columbia* and *Raleigh* had some repairs to complete but were expected to join the squadron shortly.[83] *Fern* had been attached to Bunce's command officially in May, giving the North Atlantic Squadron an organic tender and dispatch ship.[84]

The Navy Department made a special effort not to interfere with the movements of the squadron during the exercises. This meant declining many requests received by Secretary Herbert for ships to take part in various celebrations and commemorations, as they had done earlier in the year. Newspapers billed the summer exercises as the "largest fleet ever assembled by the Navy Department for instruction in fleet tactics."[85] The squadron arrived in Hampton Roads on 9 August. *Amphitrite* and *Fern* had to be detached on the way down from New York, as they were unable to keep up with the rest of the warships. *Columbia* arrived on station, overtaking the squadron on the morning of 9 August, prior to their arrival at Hampton Roads. Rear Admiral Bunce reported satisfactory progress with tactical exercises and signal drills to ascertain the range of visibility of day signals and audibility of fog signals at night.[86] Subcaliber target practice was carried out with artillery and with small arms.[87] *New York* practiced torpedo firing using her three torpedo tubes, one in the bow and two amidships. Aiming the torpedoes by eye took some practice. A typical torpedo drill involved a target being placed out and the ship steaming by the target at six, nine, and eleven knots, launching a torpedo as it passed. The speed increased with each pass.[88] There is little evidence in 1896 of the kind of fleet-level maneuvers that would later coordinate vessels utilizing torpedo attacks with battleships and their guns.[89]

On 11 August Bunce was informed that *Massachusetts* had been placed in commission and ordered to join his command. *Massachusetts,* one of three "sea-going coast-line battleships" authorized by Congress in 1890, had a main battery of four 13-inch rifles in two turrets, plus the turret-mounted secondary battery of eight 8-inch rifles and four 6-inch rifles, making her the heaviest-armed U.S. ship built to date. At 10,288 tons, she and her sister ships *Indiana* and *Oregon* were smaller than the largest battleships being built by European powers, but their heavy armament made them nominally the most powerful battleships on the ocean at the time.[90] Bunce immediately sent Captain Rodgers orders to take on his ammunition load and one thousand pounds of coal and join the squadron, either in Hampton Roads or at the squadron anchorage at Tompkinsville, where he expected to be by 25 August.[91] The rest of the squadron left Hampton Roads on 15 August to proceed back north. Before leaving the southern drill grounds, the squadron spent three days "exercising at tactical maneuvers and signals" and devoted the entire day on 19 August to target practice. This was another underway practice, with the squadron warships firing on anchored targets as they steamed past. Torpedo practice was carried out on 21 August, the ships steaming by a fixed target and launching torpedoes from their amidships tubes as they came to bear. Upon completion of these drills, the squadron turned north.[92] It arrived back at Tompkinsville on 23 August, steaming into the squadron anchorage in double columns. The outer column consisted of *Columbia, Indiana, New York,* and *Massachusetts,* while the column closer to Staten Island was *Cincinnati, Amphitrite, Raleigh,* and *Newark.* It was an impressive display of naval power, and it had a purpose. The Chinese ambassador, Earl Li Hung Chang, was in New York, and the squadron was scheduled to be inspected as part of his reception.[93] New York had been the site of impressive naval displays before. The 1893 International Naval Review, under Rear Admiral Gherardi, had served notice of the U.S. arrival as a naval power. But that fleet had been assembled under Gherardi's command specifically to execute the program of the naval review. While the Naval Review Fleet had done much formation and signaling work, it was not a combat unit, and it was never handled as such. The squadron that steamed up from Lower New York Bay in double column formation to anchor off Staten Island, and that was going to be reviewed by the Chinese ambassador, was not fashioned especially for the purpose—it was simply the North Atlantic Squadron. The next day *Texas* reported for duty with the North Atlantic Squadron by telegram. Bunce responded with orders to join the squadron at Tompkinsville as soon as ready for sea.[94]

The one contingency that caused Rear Admiral Bunce to have to detach ships was the enforcement of U.S. neutrality in the Cuban Revolution. Bunce was trying to keep the time spent in Key West, away from the squadron, to about seven to eight weeks. *Montgomery* had been on station since 30 July and so needed relief before the squadron left on their next series of maneuvers. *Newark* was designated to take *Montgomery*'s place, and Captain Stirling was ordered to have her in Key West on or about 1 September.[95] Meanwhile, with the Chinese ambassador's reception complete, Bunce made preparations to return to sea. Landing exercises were on the docket for this underway period, with Fisher's Island again the destination. Along the way *Massachusetts* would be detached to pull in to Newport for her torpedo outfit.[96] The squadron got under way on 1 September. Bunce made a point of having the four battleships, *Massachusetts, Indiana, Maine,* and *Texas,* steam together in line and column. This was a watershed moment, marking the first time a division of U.S. battleships operated together as a body. The North Atlantic Squadron had a true battle line, and Rear Admiral Bunce reported "no difficulty found whatever in their steering or handling in evolutions."[97]

The weather did not cooperate fully off Fisher's Island, and Bunce had to cancel a couple of days' operations ashore and postpone his departure for Tompkinsville an extra day, but valuable training was had nonetheless.[98] The squadron got under way from Fisher's Island on 16 September. *Massachusetts* had finished the installation of her torpedo outfit at Newport and was waiting off Block Island to join the squadron as it steamed past. The continued poor weather hampered squadron evolutions, but one clear day did give Bunce the opportunity to exercise his squadron in tactical formation, breaking the six ships up into two divisions of three ships each. They arrived at the squadron anchorage off Tompkinsville on 19 September.[99] This time they were home for ten days before heading for the southern drill grounds on 1 October.[100] From Hampton Roads, Bunce detached *Raleigh* to proceed to Southport, North Carolina, to receive the testimonial the citizens of that state had asked to present earlier in the summer. Bunce noted, "The Squadron exercises have reached a point where the *Raleigh,* an exceedingly well drilled ship, can be spared at any time."[101]

The rest of the ships worked off Hampton Roads for two weeks, returning to Tompkinsville on 14 October. It was on this voyage, while driving through a heavy gale, that *Indiana,* under the command of Capt. Robley Evans, had the locking devices on all her turrets fail, allowing the massive armored structures and the guns inside them to swing freely in the storm. In his memoirs, Rear Admiral Evans described the scene as officers and men worked frantically to secure the 13-inch guns: "I stood by the wheel on the upper bridge and frequently the whole forward end of her would go under water, men and all. . . . At such times, I held my breath as the water rolled off and the black heads of the officers and men, one after another, came in sight. I fully expected to see them swept overboard by the dozen."[102] Amazingly, no one died in the episode, although a junior officer did lose his leg when an armored watertight door broke loose from its fittings and slammed on it. By 11:10 the next morning, after a sleepless night, everything was secure. Bunce's report somewhat matter-of-factly noted that "she [*Indiana*] rejoined Squadron formation and full speed was resumed."[103] Besides the unfortunate episode with *Indiana,* which necessitated the redesign of her turret locking mechanisms, the squadron had an opportunity to exercise the new *Squadron Tactics* manual. Bunce and his staff had compiled this volume specifically for the North Atlantic Squadron, building on Commodore Parker's by-now-dated tactical manual, *Squadron Tactics Under Steam.*[104]

After testing it with the squadron during underway exercises, Bunce reported to the Navy Department that the new manual was of "great help in the discharge of my duty and if desired they can be readily perfected and issued."[105] As 1896 drew to a close with the ships of the squadron at Tompkinsville, it marked a year of the most vigorous operational exercises of U.S. warships that had ever been undertaken in peacetime.

The Blockade of Charleston

The end of 1896 found most of the North Atlantic Squadron's warships in the New York or Norfolk navy yards, undergoing upkeep and repairs.[106] While his ships were otherwise engaged, Rear Admiral Bunce was busy with the planning for the squadron's next major exercise. The series of maneuvers the squadron had carried out in 1896 had originally been conceived to proceed down the length of the East Coast. In keeping with that concept, the next exercise was set to take place in the South, off Charleston, South Carolina. This gave the added benefit of allowing citizens other than those in New England and the mid-Atlantic region to see the warships of the North Atlantic Squadron. On 8 December, Bunce received authorization to move his squadron from New York to Hampton Roads.[107] He cooperated closely with the Bureau of Navigation—something his predecessors often resented or did not care to do. Major movements of his warships were always prefaced with communications with the Navy Department, and the receipt of their blessing before he made a move. No sooner than Bunce had received the department's approval to begin deploying his forces than he received a communication from Rear Admiral Ramsay at the Bureau of Navigation. In attempting to manage the antifilibustering efforts off the Florida coast, the Navy Department had changed the orders of three of Bunce's ships, and Ramsey had communicated those changes directly to the ships without consulting Bunce.[108] Bunce took this in stride, showing his comfort with the consensus-building, managerial style of leadership demanded by the new Navy.

Rear Admiral Bunce and the representatives of the Navy Department and the Naval War College decided during their meeting that the squadron was to arrange itself in the best possible position to blockade the entrance to Charleston, while smaller vessels, such as *Vesuvius,* would be detailed to attempt to run the blockade under various conditions. The naval station at Port Royal, South Carolina, would be used to stockpile coal and supplies for the smaller vessels of the squadron, while the battleships and other vessels of larger displacement would have to coal and provision at Hampton Roads.[109] Once the particulars had been worked out and approved, Bunce returned to *New York* to carry out the preparations.

The North Atlantic Squadron, for the first time, was exercising an offensive capability. It was couched in terms that suggested that they were experimenting to assess the capability of Spain to blockade a major port on the East Coast, but there is little evidence to suggest that this explanation was any more believable in 1897 than it is today. The North Atlantic Squadron was perfecting the techniques to enable them to carry the fight to an enemy's waters, not practice coastal defense.[110] Rear Admiral Bunce's staff had to tackle several problems associated with deploying a large, disparate force away from its home waters. The battleships could not move in close to the coast to chase blockade runners, so smaller vessels would be have to be used in conjunction with the capital warships. Nor could the large ships utilize Port Royal as a depot because of their draft, so other ways would have to be found to supply and refuel them.[111] In other words, this was a protean combined arms "fleet" that required more staff work and planning than a force composed of a few ships of roughly the same size, such as Rear Admiral Walker's Squadron of Evolution, six or seven years earlier.

Bunce ordered the squadron to have steam up and be prepared to depart Hampton Roads on the morning of 3 February 1897. In issuing his orders, he broke the squadron into three sections. The first was *New York* and *Maine,* the second was *Indiana* and *Marblehead,* and the third was *Amphitrite* and *Columbia.*[112] The short trip down was marred by a violent storm, which resulted in the injury of six of *Marblehead's* men. They had to be landed at Charleston and taken to the hospital.[113] The remainder of the squadron anchored outside the harbor. On 9 February the squadron took up blockading positions, anchoring with 1,500 yards between the ships on a line of bearing northeast to southwest. The line was initially set with *Marblehead, Maine, New York, Amphitrite,* and *Columbia.* On 10 February the mayor of Charleston and the Committee of Reception and Entertainment officially welcomed the North Atlantic Squadron to Charleston and offered them the freedom of the city.[114]

With the official courtesies dispensed with, the ships immediately got down to work, practicing light discipline on the first night. The squadron's tender, *Fern,* was sent out to inspect each vessel and report to the admiral on any light that remained visible. Each commanding officer received a report on the results the next morning. They also extended the blockade line to the efficient range of the ships' searchlights to ascertain the maximum distance apart ships could be and still maintain an effective blockade in the dark, which was determined to be an interval of about three thousand yards.

The squadron closed up to four hundred yards between ships on the morning of 10 February, debriefed the previous night's work, and spent the day conducting torpedo practice.[115] That evening night target practice utilizing searchlights was carried out. Heavy weather beset the squadron on 11 February, preventing much in the way of meaningful drills. On the evening of the eleventh, the ships were joined by *Indiana,* which took her place in line, and *Vesuvius,* which was to be used as the "blockade runner." The weather was also bad on 12 February, postponing the official beginning of the exercises. *Marblehead* and *Amphitrite* went in for coal, the former preparing to depart and proceed to Florida to relieve *Dolphin* on the antifilibuster enforcement detail. The newspapers, which reported daily on the squadron's progress, were disappointed that bad weather had delayed things and eagerly awaited the commencement of *Vesuvius'* attempts to run the blockade.[116]

That evening, after sundown, *Vesuvius* got under way and headed out to sea. The line was set with four ships at three-thousand-yard intervals. Unfortunately, the fog that rolled in soon afterward led to an inauspicious start for the blockade. "The VESUVIUS had no difficulty in getting in," remarked Rear Admiral Bunce somewhat dourly in his report.[117] His brevity was more than compensated for by the newspapers the following morning. "THE BLOCKADE IS BROKEN," exclaimed the *New York Times*. The article's subheading read, "It is Demonstrated that a Blockade Runner Can Safely Pass Lines of Battleships in a Fog, Despite the Searchlights."[118] Bunce's officers complained, off the record, of course, that the fog and the fact that they only had four ships on station were all factors that contributed to an "unfair" test, but the fact remained that *Vesuvius,* with her lights doused, had steamed right in between two of the warships and reached her objective without being challenged.

The following day things went a little smoother for the squadron. *Amphitrite* returned from coaling, and at noon *Massachusetts* reported her arrival on station to the flag. After a day spent in target practice, Rear Admiral Bunce was ready to try again with six warships. The weather on the evening of 13 February was clear and moonlit, with only a slight haze. This time, things went more according to plan. *Vesuvius* made four runs and was spotted and "captured" each time before passing the blockade line. Bunce offered his assessment of the exercise as follows: "I think it has been established that, in blockading, a belt of light two miles in width, and whose length is limited only by the number of ships available, can be stretched around any harbor by the use of search lights." Bunce's remarks in his after-action report dispel any notion that this exercise was performed to try to evaluate the ability of "the Spanish" to blockade a U.S. port city. There is no mention of an enemy force anywhere in the report, Bunce's only interest being the efficiency of the North Atlantic Squadron's blockading capabilities.[119]

Rear Admiral Sicard Takes Over

Having concluded one of the largest naval exercises undertaken since the Civil War, Rear Admiral Bunce's attention returned to domestic matters. The approaching mild weather of spring brought a season of celebrations and commemorations. The presence of the Navy was highly sought after at each of these, and the Navy Department had a political minefield to negotiate in granting or denying requests. Meanwhile, a new administration took office in March. As of 6 March 1897, Bunce had a new superior. President McKinley's selection of John D. Long as his secretary of the navy was a fortunate choice for the New Steel Navy. Previously a governor of Massachusetts as well as a member of Congress for six years, Long had contacts on Capitol Hill, as well as ties to an important maritime state, that made him an effective secretary. But it was McKinley's selection a month later of a young New Yorker, Theodore Roosevelt, as the assistant secretary of the navy that would have an even more historic impact. While Roosevelt rightly gets much of the credit for preparing the U.S. Navy to prosecute the War of 1898, this study has demonstrated that the North Atlantic Squadron, under Rear Admiral Bunce's energetic leadership, spent much of the year prior to Roosevelt's appointment as assistant secretary busily engaged in developing its skills.

The squadron (except the monitors) left Charleston on 21 February. *New York, Indiana,* and *Columbia* proceeded back to Hampton Roads, where *Raleigh* awaited them. *Massachusetts* went on to New York to enter the navy yard and test the new dock that had just finished construction. From there she was to go to Boston where the citizens of her namesake state were to present her with a memorial.[120] *Newark* was in dock at Port Royal. *Texas* and *Maine* were detailed to New Orleans, with orders to arrive in time to participate in the Mardi Gras celebration, and *Montgomery* left her station in Key West to be present at the Mardi Gras festivities in Mobile, Alabama. *Marblehead* and *Vesuvius* both remained in Florida to reinforce the expanded antifilibustering mission.[121]

The fact that the Caribbean was relatively quiet in the winter of 1897 helped the squadron have enough ships available to carry out multiship exercises. The absence of urgent threats to U.S. lives and property allowed the Navy Department to use the Apprentice Training Squadron ship *Essex* to pay port visits to many of the usual stops of squadron warships in the West Indies. Essex's spring cruise in 1897 included stops in Barbados; La Guayra, Venezuela; and Kingston, Jamaica. This not only provided valuable training for the apprentices in *Essex* but also was a low-cost alternative to issuing individual orders to squadron warships, thereby allowing Bunce to keep his unit concentrated, and was much less threatening to the Spanish than sending the entire squadron.[122]

On 6 March 1897, the squadron got under way for Tompkinsville. *New York* was to perform a full-speed trial on the way home, so Bunce arranged for the squadron to break up and proceed independently after clearing the Virginia Capes.[123] It was one of the rare times during Bunce's command that the flagship moved without having tactical control of at least one other member of the squadron—a far cry from the year 1873, when no squadron ships sailed in company at any time during the year. Bunce, in fact, was constantly concerned with operating his warships at least in pairs whenever possible, the better to exercise their station-keeping abilities. His instructions to *Texas* and *Maine,* returning to New York from their Mardi Gras duty in New Orleans, were explicit that the two ships would be exercised during the transit. "Your attention is called to the Programme of Exercise for Section, issued by me June 15th 1896, a copy enclosed," he wrote to Captain Glass.[124] Likewise, even the monitors—although they did not exercise with the squadron—were sent out in section whenever possible.[125]

The success of the naval review was a fitting close to the admiral's tenure as commander in chief. On 1 May Bunce detached from command of the North Atlantic Squadron and reported as the commandant, New York Navy Yard and Station. Although he never took the squadron on an operational deployment, Rear Admiral Bunce can be credited with coming the closest yet to realizing the transformation that Rear Admiral Luce had envisioned for the squadron a decade earlier. During Bunce's tenure, single-ship deployments were kept to a minimum, while the entire squadron, now reinforced by the arrival of battleships, conducted two major and several lesser exercises. In Adm. Robley Evans' words, "We had mastered it [handling battleships] in the only way possible to seamen—by constant work and practice out on the blue water. We all owe much to Admiral Bunce."[126]

Rear Admiral Bunce was relieved by Rear Admiral Montgomery Sicard, whose previous assignment had been commandant of the New York Navy Yard.[127] Like all the admirals of his generation, Sicard had a distinguished Civil War record, having been present at the capture of New Orleans in 1862, run the batteries at Vicksburg in 1863, and commanded the gunboat *Seneca* during the assaults on Fort Fisher in 1864–63. Postwar duty included tours at the United States Naval Academy, various sea commands, and nine years as chief of the Bureau of Ordnance.[128] Sicard inherited a squadron focused, at least in the short term, on domestic matters. However, greater problems were brewing in the North Atlantic Squadron's area of operations.

The longer the civil war in Cuba dragged on, the more desperate the Spanish Empire became to suppress the insurrection. By 1896 the Spanish forces had burned entire villages and planted fields, seized livestock, and rounded up Cuban civilians to be placed in the infamous "reconcentration" camps.[129] All of this, of course, was bad for U.S. business interests on the island, which amounted to $50 million in direct investments and another $100 million in trade.[130] This, coupled with the vibrant expatriot Cuba Libre movement, contributed to a constant interest by various combinations of U.S. citizens and Cubans to become involved in helping the rebels eject the Spanish from the island. As has already been shown, throughout 1896 and 1897, attempts to maintain U.S. neutrality were leading to an increasing workload for the squadron. What had begun as an assignment of one vessel to duty off Florida had become a constant deployment of three North Atlantic Squadron warships by the time Sicard took over. The trouble brewing in Cuba pushed Sicard to continue drilling his Squadron, preparing for whatever crisis might develop.[131] The department took a renewed interest in training and preparedness as well, detaching *Amphitrite* from the squadron on 7 May to serve as a training ship for gun captains.[132]

Spring began the season for celebrations and commemorations. In the midst of servicing all these requests from citizens that wanted to either see their warships or have them actually protect them, Sicard became concerned that the cohesion of his squadron and its new mission to maintain the ability to fight a fleet action might suffer. In June he wrote to the secretary of the navy, expressing emphatically the importance of keeping "together as many of the vessels of this squadron as the demands of the service will allow." Sicard went on to note: "Being 'in squadron' is of great advantage to the discipline and order of ships, as it promotes emulation between their officers and crews, and the frequent signaling made necessary by the presence of numerous vessels, keeps the personnel watchful, alert, and attentive, and accustoms them to constant use of the different kinds of signals—in other words the squadron is an excellent school of practical, every day duty."[133] Sicard, like Bunce and Luce before him, envisioned the North Atlantic Squadron as a fighting organization constantly devoted to training, not an administrative grouping of ships.

Rear Admiral Sicard's concerns were answered by Assistant Secretary of the Navy Theodore Roosevelt. In a letter sent just two days after Sicard mailed his from Hampton Roads, Roosevelt reassured the admiral, stating that "the Department desires to keep the squadron intact after August 1st, and from that date you will have ample opportunity for squadron drills."[134] At the time of his correspondence with Sicard, Roosevelt had just returned from Newport, Rhode Island, where on 2 June 1897 he had given one of the pivotal speeches of his career to the Naval War College class of 1897. "We must therefore make up our minds once for all to the fact that it is too late to make ready for war when the fight has once begun," he had said.[135] Sicard had a true ally in Roosevelt, who was eagerly in favor of keeping the squadron together as much as possible and understood that preparedness required constant training and learning. Two weeks later Sicard traveled from Norfolk to Washington, D.C., where he met personally with Roosevelt as well as Cdr. Goodrich of the Naval War College. Together with Captain Crowninshield, chief of the Bureau of Navigation, the three men planned the squadron's August maneuvers.[136]

After a week in Bar Harbor, Maine, where the commander in chief and his officers partook of the usual festivities, the squadron got under way on 30 August for the southern drill grounds off Hampton Roads and Admiral Sicard's long-awaited summer maneuvers.[137] These maneuvers were closely observed by

Assistant Secretary Roosevelt, who joined the squadron in *Dolphin* on 7 September. Roosevelt spent two full days with the squadron, inspecting *Iowa* and *Brooklyn,* observing target practice—both service and subcaliber—searchlight drills, and squadron evolutions, and departing on 9 September. The maneuvers continued for another three days, finishing on 12 September. Sicard pronounced the results "generally . . . satisfactory."[138]

After spending a couple of weeks coaling the ships and performing routine maintenance, Sicard wrote to the department with his plans for the next month. He envisioned a series of maneuvers, taking the squadron first to the southern drill grounds from 27 September to 1 or 2 October, then to the Yorktown area where the naval brigade could be landed for drills and target practice. After recoaling, the squadron would be ready for whatever other duties the department might have in mind.[139] What followed was a bizarre exchange of letters with Assistant Secretary Roosevelt that speaks volumes about the change in the character of fleet command in the 1890s.

Roosevelt's leadership within the Navy Department during Secretary Long's lengthy illnesses and visits to his home is a well-known fact. The story of Roosevelt's telegram to Dewey in Hong Kong, telling him to coal his ships and attack the Philippines in the event of war with Spain, is legend among even the most casual students of the war. What is less known is the extent to which the hyperactive Roosevelt was deeply involved in the operations of the Home Squadron during this time. On 21 September Sicard received this letter from Roosevelt, acting as secretary in Long's absence: "The Department suggests that advantage be taken of all the passages of the squadron under your command from one port to another to engage in fleet maneuvers, instead of waiting until your arrival upon the regular drill grounds in order to engage in these exercises."[140] One can only imagine how Rear Admiral Walker or Rear Admiral Meade would have received this letter. Previous chapters have documented Walker's multiple-page missives fired off for much less provocation than this. Sicard, however, remained calm. In a respectful reply, he gently instructed Roosevelt in how units of multiple ships move together and what sort of environmental factors govern their movements then went on to say sympathetically, "I am fully alive to the importance of practicing evolutions whenever opportunity in afforded . . . and shall do so."[141]

The squadron anchored off Yorktown, Virginia, on 27 September, Sicard having decided to do the landing drills and ashore work prior to taking on coal. The planned rifle target practice for the sailors had to be postponed until the squadron could find another location for a makeshift range. The Yorktown area had become so settled that Sicard was concerned that the rounds from the Navy's new 6-mm rifle would endanger civilian lives or property, so he decided to await another opportunity elsewhere. The squadron made up for the lack of rifle target practice with a week of "exercising landing brigades on shore, in extended movements by companies, and in target practice with pistols and revolvers." On 1 October sailors from *New York, Maine,* and *Puritan* set up defensive positions on Gloucester Point. They were then attacked by the battalions from *Brooklyn, Iowa, Massachusetts,* and *Indiana.* After a day of fighting, the offense was judged to have won the skirmish. Upon completion of the landing exercises, *Fern* left for Norfolk, Virginia, with a draft of handpicked men to be trained as gun captains on board *Amphitrite,* which had been stationed at Norfolk since being detached from the squadron in May.[142] The rest of the squadron departed Yorktown on 4 October, headed to the southern drill ground for maneuvers. *Brooklyn* was detached to Hampton Roads when one of her main steam pipes began leaking and required repairs. *Puritan* was detached to proceed to the New York Navy Yard for scheduled repairs. The remaining ships—*New York, Iowa, Massachusetts, Indiana,* and *Maine*—conducted tactical maneuvers from 5 to 9 October.[143] *Indiana* and *Maine* were detached by the Navy Department to proceed on other business. The three remaining ships then proceeded to Cape Cod Bay, where they met up with *Texas* and conducted another set of tactical maneuvers on 13 October before departing the area for Boston, where the squadron participated in ceremonies marking the one hundredth anniversary of the launching of the USS *Constitution* on 21 October.[144]

The squadron arrived back at Tompkinsville on the twenty-fourth of October. *New York* went into dock immediately for a scheduled upkeep period.[145] Although not under way, Admiral Sicard never lost sight of the vital importance of continuing with training and preparedness. He dispatched *Texas,* which had been in the New York Navy Yard for much of the previous underway period, to the northern drill grounds for four days of target practice on all guns, following a set of instructions very carefully written and issued by him. Captain Wise was admonished to "express [his] opinion of the practice, using for that purpose, the Naval Academy scale of merit."[146] It is a myth that accuracy with the great guns was ignored until after the War of 1898. Although there were charges of "gun-decking" target practice scores, and the real-world postbattle analysis of the accuracy at the Battle of Santiago de Cuba showed an abysmal hit rate, the fact was that better accuracy awaited advances in technology, not more effort on the part of the squadron or her commanders in chief to practice with the guns.[147] During this time, Sicard also issued a number of orders to squadron warships concerning the ongoing efforts to patrol the Florida coastline. *Montgomery* and *Vesuvius* were ordered to Pensacola and Jacksonville, respectively, to join *Detroit,* which was under the Navy Department's direct jurisdiction patrolling the Key West district.[148]

The squadron now regularly not only went to sea together but also spent time not at sea in the same port, usually Tompkinsville. Time in port together inevitably meant that the crews of the warships would spend their leisure time together. The time-honored tradition of racing the ships' cutters has already been mentioned. Previous chapters have touched on the existence of baseball teams on board some of the larger ships.[149] The 1890s craze for sporting and other leisure activities carried over to the Navy, with other sports soon joining rowing and baseball. It is important to note that these sports were organized and enjoyed on a squadron level, not by ship. Bicycling was becoming popular during this time, and two of the squadron's commanding officers, Captain Silas Casey of *New York,* and Captain Francis Higginson of *Massachusetts* were avid cyclists. A cycling club started on *New York,* but that was insufficient to serve the numbers of officers who were interested. Soon a circular was sent to the wardroom of every ship in the squadron, and the North Atlantic Squadron Bicycle Club was born, complete with elected officers and representatives from each warship. At this point the officers not only were comfortable with one another when handling their ships in formation and mingled with each other at official functions and receptions, but they now formed relationships and shared interests to the point where they were participating in non-official recreational organizations on a squadron level.

Conclusions

Through 1897 the trouble in Cuba escalated to the point that the Navy Department eventually wanted the entire North Atlantic Squadron in Florida. Sicard was alerted on 3 December that the department desired the squadron's winter cruise and exercises to take place off Key West and the Dry Tortugas. He was directed to concentrate his squadron at Hampton Roads prior to moving south.[150] In compliance with this plan, Sicard began to issue orders to his warships. Throughout 1897 Sicard had worked to successfully prepare his squadron for the deployment they were now undertaking.[151] The flagship moved to Hampton Roads in December. In his instructions to the other warships that would be joining the squadron when their repairs were complete, Sicard was emphatic that all line officers be afforded an opportunity to practice the tactical maneuvers they would be expected to be familiar with. He even enclosed a copy of a table that had been created on *New York* containing each junior officer's name, the various evolutions practiced, and the date that officer acted as the officer of the deck during one of the evolutions. He strongly suggested that his other warships take up this practice of systematic training.[152]

On 8 December, Bunce ordered Captain Sigsbee of the *Maine* to "proceed with the MAINE under your command to Key West, Florida, and there await further orders."[153] These were fateful instructions. In his memoirs Rear Adm. Robley Evans described those last few weeks of *Maine's* short existence: "At Key West, Florida, I found the North Atlantic Fleet under the command of Rear Admiral Sicard, and it was clear to me that that able officer expected war with Spain and was doing all he could to be ready for it when it came."[154]

The previous chapter ended with a commander in chief who was unsure of the locations or assignments of any of his warships, other than the flagship. That was in 1894. Three years later the squadron not only traveled in formation but also trained together, using standardized training guides promulgated by Rear Admiral Sicard. The crewmembers socialized together after hours, engaging in activities such as the North Atlantic Squadron Bicycle Club. In the three years prior to the War of 1898, the commander in chief of the North Atlantic Squadron was in direct tactical control of at least one of his ships for 536 of 1,095 days, or 49 percent of the time. During this time, ten major squadron exercises took place. These facts represent a major change in both mission and identity. The squadron's primary mission shifted from single-ship "showing-the-flag" and presence operations to being prepared to confront an enemy fleet. With that change in function came an identity as a tactical unit that had not existed fifteen, or even three, years earlier.

Significant advances were made in multiship operations. Under the leadership of Rear Admiral Meade, the "usual winter's cruise" in 1895 was accomplished not by sending the warships out piecemeal to various ports in the Caribbean but by moving as a squadron throughout the operating area. Although Meade's attitude and leadership style were incompatible with the demands placed upon a modern commander in chief, he was effective in keeping the squadron concentrated and developing its unit identity. Under Rear Admiral Bunce, the squadron reached new heights in the development of doctrine and tactics through a series of meticulously planned exercises. The support and development of a naval militia enabled the squadron to restrict its involvement in harbor protection. With the monitors in ordinary manned by willing naval militia volunteers, the North Atlantic Squadron could focus on the tasks required of a seagoing battle fleet.

There were still challenges to be met before the true power projection capabilities of the organization could be realized. The purchase and delivery of high-quality coal for so many warships concentrated in one place continued to consume an inordinate amount of time for both the squadron and the Bureau of Equipment and Recruiting. The steel ships had to be docked more often, and the lack of availability of suitable dry docks—especially for the larger ships—was a limiting factor. Target practice was unrealistic, and the concept of conducting exercises against opposing forces had not yet been perfected. Still, at 1897's end, the North Atlantic Squadron was a changed organization. Consisting of entirely new matériel and better trained, it entered 1898 as a coherent combat unit, prepared for combat as a squadron.

Epilogue

It is appropriate that this study end in the same geographic location where it began twenty-three years earlier: the waters off the coast of Key West, Florida. There, at anchorage off the island of Dry Tortugas, the North Atlantic Squadron, with Rear Adm. Montgomery Sicard in command, was resting after a day of exercises. Late in the evening of 15 February 1898, a torpedo boat from the naval station came alongside the flagship *New York* with shocking news of the explosion of the battleship *Maine* in Havana, Cuba. The much-anticipated war with Spain was now almost a certainty. The North Atlantic Squadron was ready, not simply because it possessed modern steel warships but also because it had spent the last two decades engaged in an ongoing process of transformation from an administrative unit to a group of warships constituting a battle fleet. The process was not complete in 1898, but enough progress had been made to test the North Atlantic Squadron's ability to engage another naval power in a multiship action, should the need arise.

A full scholarly treatment of the North Atlantic Squadron's actions during the War of 1898 is beyond the scope of this study. However, a brief review of the squadron's accomplishments between February and July 1898 will serve as a background upon which to consider the arguments of this book.[1] This study has offered evidence that the creation of fleet practices crucial for the sound deployment of the modern warships of the New Steel Navy began in the 1870s, grew in the 1880s, and matured in the 1890s. The result of this ongoing process was the construction of an organizational identity as a combat unit.

Swinging at anchor with Sicard's flagship, the armored cruiser *New York,* were the battleships *Iowa, Texas, Massachusetts,* and *Indiana*. During the 1874 crisis, the order for a concentration of the fleet in Key West was made during a meeting of President Grant's cabinet on 14 November 1873, and it was not until the end of January 1874, after the threat of hostilities had already passed, that enough of a "fleet" could be mustered to hold tactical exercises. By late 1897, enough new ships existed that the North Atlantic Squadron could remain concentrated as the norm rather than as the exception.

After the war Rear Admiral Sicard was criticized for his first actions after finding out *Maine* had been destroyed. Some felt that Sicard should have weighed anchor and got under way with the battleships for Havana instantly. Perhaps the show of force of his battleships in Havana Harbor would have caused the Spanish to capitulate and might have averted hostilities. Rear Adm. John G. Walker or Rear Adm. Richard W. Meade might have done just that. As has been shown, these officers' concept of squadron command took an expansive view of the powers of the commander to do as he thought best and seek permission later. Sicard reacted as the modern institution would have expected him to. He moved *New York* closer to Key West that evening to stay in closer telegraphic communication with Washington, D.C., and awaited orders from the Navy Department concerning the disposition of his squadron. Sicard's actions are evidence that the mode of leadership of the commander was undergoing a process of change as well—from what Morris Janowitz has called a "heroic" leader to a "managerial" one. Sicard regarded his command not as his personal fiefdom but as an instrument of power to be directed by Washington, D.C.[2]

Rear Admiral Sicard was replaced in command of the North Atlantic Squadron by Capt. W. T. Sampson, *Iowa*'s commanding officer, on 28 March 1898. Like Sicard, Sampson had been assigned previously as the chief of the Bureau of Ordnance, prior to assuming command of *Iowa*. Much has been made of Sicard's relief by Sampson. The accusations are twofold: Sicard had somehow been found lacking by Secretary of the Navy Long and Sampson was advanced over the heads of several officers senior to him due to some kind of favoritism. Both charges are vehemently denied by Secretary Long in his memoirs.[3] The facts are simply that a shooting war was coming and Sicard was demonstrably unwell. Secretary of the Navy Long (not exactly healthy himself) prudently had to get someone into the commander in chief's position who was physically capable of the demanding duties associated with leading a squadron of warships in a fleet combat action. Sampson, although only a captain at the time, was the next senior officer in the North Atlantic Squadron, and it was his rightful place to take the squadron in the event of his senior's incapacity. Sampson had the added advantage of having worked with the squadron while captain of *Iowa*. It was hardly the right time to attempt to break in a flag officer who might have been senior to Sampson but was serving ashore when the crisis erupted.

The organization of Sampson's fleet soon became an issue for the Navy Department. Faced with the threat of an actual war, the idea of a concentrated fleet broke down in the face of popular clamor for coast defense. The extensive public relations work done by the squadron, documented in this study, produced an unexpected side effect. Immediately upon the threat of hostilities, cities up and down the East Coast demanded protection from the threat of the unlocated Spanish squadron. The Navy Department's first response was to mobilize thirteen Civil War–era monitors and man them with personnel drawn from the ranks of the naval militia. However, even civilians understood that the ancient monitors and their ridiculous smoothbore artillery were no protection against the capabilities the Spanish armored cruisers possessed.

Eventually the decision was made to constitute two combat squadrons and one patrol squadron with all the assets available. In addition to Sampson's North Atlantic Squadron, the so-called Flying Squadron was formed at Hampton Roads, which was believed to be close enough to Key West to reinforce the North Atlantic Squadron if necessary, yet close enough to New York to protect it if the occasion arose. *Texas, Massachusetts,* and *Iowa* formed its nucleus, with *Brooklyn* as the flagship of Rear Admiral Schley. Farther north, the Northern Patrol Squadron, commanded by Rear Adm. J. A. Howell, was built around a core of four converted Morgan Line steamships, renamed *Yankee, Dixie, Prarie,* and *Yosemite.* These ships were scattered up and down the East Coast, ostensibly to provide early warning of the approach of the unlocated Spanish squadron. The real purpose of the vessels was to assuage the fears of the public, which were greatly magnified by the constant alarmist reporting of newspapers. The Northern Patrol Squadron served its purpose. If not for the naval militia, the assets available to the combatant squadrons would probably have been drawn down even further.

Upon the declaration of war by Congress on 21 April, the squadron was immediately ordered to get under way and establish a blockade of major Cuban ports. Early on the morning of 22 April, *New York* departed Key West, followed by *Indiana, Cincinnati, Wilmington, Helena, Machias, Nashville, Castine,* seven torpedo boats, and three monitors. The challenge facing Sampson was to mount an effective and maintainable blockade that would deprive Spanish troops of the supplies necessary to continue fighting while keeping in mind that the whereabouts of a Spanish squadron that was said to have left Spain under the command of Admiral Cervera were not known.

This latter complication was the reason that the Navy Department would not allow Sampson to close on Havana and shell it into submission with his battleships, as he had requested permission to do. The plan might have worked, but the Navy Department objected that there were no infantry forces available to occupy the city, even if it did surrender. Moreover, after the shocking loss of *Maine,* public opinion would not stand the loss or serious damage of another capital warship. Sampson was required to preserve his armored ships for an anticipated battle with a Spanish fleet.

After receiving word that Cervera had left the Cape Verde Islands on 29 April, Sampson left a few cruisers off Cuba and took the rest of the squadron to San Juan, Puerto Rico. His arrival was delayed by the fact that he had to have the two monitors assigned to the squadron towed. It did not matter, however,

because Cervera was not at San Juan. The squadron bombarded the city fortifications for about an hour, with little effect, then returned to Key West to refuel.
Upon arrival there, Sampson was met with the news that the Spanish had been
sighted at Martinique. With proof that Cervera did not, therefore, pose a threat
to the East Coast, Schley's Flying Squadron was detached from Hampton Roads
and ordered to report to Sampson at Key West. Both squadrons met at the naval
station on 18 May. While they were deciding their next move, Cervera was able
to refuel at Curacao, then steam quickly across the Caribbean to the safety of
Santiago de Cuba.

While Sampson was busy hunting for Cervera, the Asiatic Squadron, under
Commo. George Dewey, delivered a crushing defeat to the Spanish squadron
guarding the Philippines at the Battle of Manila Bay on 1 May 1898. While the
Asiatic Squadron's experience is not a direct concern of this study, a couple of
observations are in order. Dewey's squadron consisted of the cruisers *Olympia*
(flag), *Baltimore, Boston,* and *Raleigh;* the gunboats *Petrel* and *Concord;* and the revenue cutter *McCulloch.* Of these seven warships, *Baltimore, Raleigh,* and *Petrel* had
served with the North Atlantic Squadron, and *Boston* and *Concord* had served
with Walker's Squadron of Evolution, prior to their assignment to a foreign
cruising station. Only the flagship *Olympia* had not had the opportunity to be
immersed in daily, multiship operations. In this can be seen evidence of the fru-
ition of Luce's vision of using the North Atlantic Squadron as a school of practical application for the fleet.

Unlike the Battle of Santiago, which will be discussed shortly, the Battle of
Manila Bay was actually fought in formation. Prior to the engagement, Dewey
formed his squadron into a column. He then directed the movements of the
column as it attacked, withdrew, and then reattacked the Spanish line. The simple column was not a complicated formation, nor did it have to be maneuvered
in a complex manner, mostly due to the fact that the Spanish squadron was stationary. Nonetheless, it can be argued that at least some of the credit for the success enjoyed by the Asiatic Squadron belongs to the rigorous exercise at sea that
six of its seven members received while with the Home Squadron/Squadron of
Evolution in the 1890s.

Meanwhile, not knowing that Cervera had steamed for Santiago, Sampson
assumed that he would be headed either for Havana or Cienfuegos, a harbor
on the southern coast of Cuba that was connected by railway with the capital. Accordingly, Sampson augmented the Flying Squadron with *Iowa* and sent
Schley to blockade the harbor of Cienfuegos. Meanwhile, he took the remainder of the North Atlantic Squadron and blockaded Havana. It was at this point
that Schley committed the first of several missteps that contributed to the acri-
mony between the two men after the war. Taking an inordinate amount of time
to reach Cienfuegos, Schley did not arrive off the harbor until 22 May, and then

he was unable to ascertain definitively whether or not the Spanish were there. Sampson, meanwhile, had received intelligence that Cervera was at Santiago and frantically sent a dispatch vessel to Schley with orders to proceed to Santiago at once and find the Spanish.

Schley kept his own counsel, however, and delayed two days in leaving Cienfuegos. When he finally got within about thirty miles of Santiago, he decided that his squadron was low on coal. Unable to refuel at sea due to bad weather, he turned the Flying Squadron around and steamed for Key West. While headed west, the squadron was met by another dispatch vessel with orders from the Navy Department to return to Santiago to find out whether Cervera's squadron was actually there and to "take appropriate action" if it was. Schley's response was "Much to be regretted, cannot obey orders," and he continued for Key West. Secretary of the Navy Long later remarked that the department could be justly criticized for not relieving Schley "at once."[4] Fortunately for Schley's career, as well as the U.S. war effort, there was a break in the weather and Schley was finally able to refuel his ships lowest on coal. On the evening of 28 May, he returned to Santiago and found the Spanish squadron in the harbor.

Sampson brought the North Atlantic Squadron to Santiago immediately. With the two forces unified, he established a close blockade of Santiago de Cuba. On 10 June he ordered a battalion of Marines ashore at nearby Guantanamo Bay to establish a coaling and resupply station. The successful operation validated the many landing exercises the North Atlantic Squadron had carried out over the previous years. Sampson then called upon the U.S. Army to land and capture Santiago's fortifications so that he could approach the harbor without exposing his capital ships to fire from the Spanish guns. In response, 16,000 men led by Gen. William R. Shafter embarked in transports at Tampa, Florida, and sailed to the southeastern coast of Cuba. The thirty-two transport ships were convoyed by Navy warships, in an escort formation reminiscent of one that had been practiced by Rear Admiral Case in 1874.

General Shafter's forces landed twenty miles east of Santiago. Sampson met with Shafter on 20 June, and on 1 July, U.S. forces attacked Santiago's outer defenses at El Caney and San Juan Hill. The Americans carried the defenses, but at a cost of over 1,500 men killed or wounded. Shafter was concerned that he would have to withdraw, and he sent a message to Sampson begging him to attack the harbor to relieve pressure on the Army forces.

This, of course, was impossible, due to the mines and the guns of the still-uncaptured Spanish fortifications. On the morning of 3 July, Sampson set out in *New York* to meet with Shafter and tell him so. Schley was left in charge of the blockade. At 9:30 a.m., the Spanish squadron sortied from Santiago. Every U.S. ship that could get up enough steam took off in pursuit of the Spanish ships, firing wildly. *New York,* with Sampson on board, turned around and frantically raced toward the scene of the battle, only to find the action largely over by the time she could get within the range of her guns.

During these initial moments, Schley's actions again came into question. Concerned either about being within range of Spanish torpedoes or that Cervera's flagship, *Infanta Maria Theresa,* was trying to ram him, or some combination of both, Schley ordered *Brooklyn*'s captain to turn her to starboard[5] in a circle, which nearly caused a collision with *Texas* and *Oregon. Texas* was required to back her engines hard, bringing the battleship almost to a complete halt, and forced to stop firing as *Brooklyn* masked her guns. *Oregon* managed to evade both *Texas* and *Brooklyn* and continue past to the east. This episode would later be referred to as "the Loop." The final outcome of the engagement, never really in question, took a little over three hours and no fewer than 10,000 shells to effect. In the end "a materially superior U.S. fleet had virtually destroyed its Spanish opponent while suffering almost no damage to itself."[6]

In the battle's aftermath, controversy erupted over which admiral deserved credit for the victory. Rear Admiral Sampson, although he had left the area for his meeting with General Shafter, remained in overall command of the victorious fleet and left no doubt in his initial telegrams to Washington, D.C., that he considered himself the victor. Rear Admiral Schley, as the acting CinC in Sampson's absence, was in command during the battle itself. However, as previously noted, his actions as officer in tactical command largely consisted of flying a signal ordering the fleet to chase down the Spanish ships. The uneasy relationship between the two admirals was at least cordial in the beginning. Schley's after-action report to Sampson congratulated him "most sincerely upon this great victory to the *squadron under your command* [emphasis added] . . . and I am glad that I had an opportunity to contribute in the least to a victory that seems big enough for all of us."[7] After the war, tension between the two continued to grow until certain remarks, attributed to naval officers, were made public that cast aspersions on Schley's conduct in the early days of the campaign and the inept way in which *Brooklyn* was handled during the Santiago action. Schley demanded, and received, a court of inquiry into his conduct in 1901. Alerted in advance that the board's findings were largely unfavorable to him, he retired quietly before it could issue its report.

Conclusions

We have no way of knowing how the North Atlantic Squadron's transformation would have been judged if it had been fully tested in 1898. Fortunately the squadron did not face a peer competitor in close-order fleet combat. The fact that the hapless Spanish ships were destroyed by a superior North Atlantic Squadron has dominated the historical narrative of the 3 July 1898 engagement, but the Battle of Santiago was not a fleet action. After Rear Admiral Schley's flagship hoisted the signals to "Follow the flag," and "Close up," the subsequent engagement resembled nothing so much as a target practice with the targets in motion. Theodore Roosevelt later referred to it as a "captain's fight," the implication being that further coordination and leadership from the flag was not required.[8]

Viewed properly, the Battle of Santiago represents not a conclusion or a finalized outcome but a waypoint in the process of the development of a U.S. battle fleet. The endpoint of this transformation requires further study. It can be argued that the first true test of American ability to fight in close order was the deployment of Battleship Division Nine to join the Royal Navy Grand Fleet during World War I. British officers were underwhelmed with the tactical abilities of the U.S. battleships, particularly with respect to gunnery.[9] Having thus been exposed directly to the best practices of a mature fleet, however, the U.S. Navy was able to develop its own battle fleet in the interwar period. The proper endpoint of a full study of the development of a U.S. battle fleet would therefore include the fleet problems of the 1920s and 1930s and end with the outbreak of World War II.[10]

This is not to denigrate the significance of the process the squadron went through to develop the ability to carry out the 1898 operations. It was, in fact, a process that deserves careful study. This work has argued that beginning in 1874, the North Atlantic Squadron underwent a slow transformation from a largely administrative organization to a coherent combat unit. It was not a linear process, but one in which progress in critical areas was modulated by conflicting demands that caused distraction. From 1874 to 1897, the squadron was constantly required to balance the missions of cruising, domestic security, and public relations with the Navy Department's desire to train for fleet combat.

The operational record narrated in this study, over a period of twenty-three years, suggests three distinct periods of ongoing development of the identity of the North Atlantic Squadron as a combat unit. This development was not continuously progressive but was interspersed with setbacks. The conduct of foreign policy and limitations of matériel interfered with the ability of the squadron to concentrate its forces in a manner conducive to conducting fleet tactical exercises and building unit identity. The squadron began conducting tactical exercises with the wooden cruising assets available. A protean concept of multiship operations was developed, yet the squadron did not possess a fleet warfighting capability, a fact widely acknowledged by naval officers.[11] An interim period began when the last of the wooden cruising vessels was replaced by steel warships. The Navy accelerated the development of multiship capabilities with the deployments of the Squadron of Evolution, the Squadron for Special Service, and the Naval Review Fleet, the latter two units concerned mainly with appearance rather than warfighting capability. While largely for public relations purposes, their deployments forced the Navy Department to deal with a range of logistical issues and provided valuable experience for junior officers in the art of close order, multiship operations. A mature—I do not say complete—capability arrived when the squadron was able to exercise its modern assets under simulated combat conditions, such as the mock blockade of Charleston, South Carolina. These exercises proved valuable when the squadron was called upon to carry out similar tasks during the War of 1898.

The U.S. Navy circa 1874 has been criticized as ineffectual. In fact, as historian Lance Buhl has argued, the United States possessed an adequate Navy given the nation's priorities in the 1870s.[12] It was highly professionalized, with entry into the officer corps controlled by the U.S. Naval Academy and the U.S. Naval Institute as a forum for professional discourse and development. From a matériel standpoint, the wooden cruising vessels in use possessed the requisite amount of firepower to successfully intimidate small Caribbean nations, while their sails allowed them to cruise on station for long periods of time economically. Fleet tactical training prior to 1874 was largely theoretical. Visionary officers such as Commo. Foxhall Parker wrote textbooks and developed signaling capabilities, and Naval Academy cadets were examined on the theory of tactical formations at Annapolis. No operational capability existed, however, and none was thought necessary until national strategic priorities changed. The *Virginius* incident in late 1873 highlighted the fact that the U.S. government could no longer pursue its strategic priorities in the Caribbean region with the Navy as it was then constituted and pressed for change.[13] A new strategic purpose for the Navy required not only new matériel but also a new concept of operations and a new identity for the Home Squadron.

From 1874 on, the Navy Department made a conscious decision to train its personnel in the mechanics of fleet tactics under steam. This was to be practical training to complement the instruction naval cadets received at the academy. The 1874 exercises were conducted at a speed of four knots, useless for an actual engagement against an enemy fleet. By 1882 the speed of the maneuvers had been raised to six knots. Tactical exercises with wooden cruising vessels had limits, though. By 1882 Rear Admiral Cooper had let the Navy Department know in several of his reports that he felt his squadron had accomplished what it could in terms of practicing tactical formations. The strategic purpose for multiship operations was still unclear, as contingencies regularly interfered with the ability to concentrate enough warships to conduct fleet tactical training. As an example, Rear Admiral Cooper's relief, Rear Admiral Jouett, spent a large portion of his commander in chief tour responding to the revolution in Colombia and the threats to the passage of the Panamanian isthmus.

Rear Adm. Stephen B. Luce wanted more. After establishing the Naval War College in 1884, he attempted to marry the theoretical work of the college staff and students with practical application by the North Atlantic Squadron warships during summer exercises. During his tour as commander in chief, the Navy Department's strategic priorities had not yet developed to this point where enough assets could be made available to carry out Luce's vision. The requirement to provide presence throughout the Caribbean and the Canadian fisheries consistently interfered with and eventually precluded his ability to carry out tactical exercises, much to Luce's frustration.

This limitation was addressed during 1889–91 by providing a venue for multiship operations that would not be subject to the requirements of station cruising. As the first four steel warships authorized in 1883 were commissioned and joined the fleet, the Navy Department made a conscious decision to concentrate them rather than assign them piecemeal to the various cruising stations. The establishment of the Squadron of Evolution in 1889, under the leadership of Rear Adm. John G. Walker, not only provided intensive experience in daily multiship operations but also required the Navy Department to address logistical issues associated with the overseas deployment of concentrated forces of warships. Rear Admiral Walker's correspondence made it clear that he viewed his primary duty as a commander in chief to be the preparation of his squadron for combat. This marked a clear departure from the concept of command and the command experience of previous squadron commanders.

The absence of an operational chain of command hindered developments during this time. The admiral of the Navy was an advisor to the secretary of the navy without executive functions. Thus individual squadron commanders did not report to a superior naval officer, but directly to the appointed secretary of the navy or his civilian assistant. This led to conflicts of personality among squadron commanders, most notably the tense relationship between Rear Admiral Gherardi of the North Atlantic Squadron and Acting Rear Admiral Walker of the Squadron of Evolution. No structure was in place to mitigate these difficulties. Walker was an advocate of Stephen B. Luce's ideas, having used his influential position as chief of the Bureau of Navigation to support Luce's efforts to establish the Naval War College. However, his personal behavior and his conflict with Gherardi precluded what could otherwise have been an opportunity for meaningful multisquadron tactical exercises in 1891–92. Indeed, the evidence shows that the Navy Department went out of its way to keep the warships of the two squadrons apart during this time. The Sampson-Schley controversy following the Battle of Santiago indicates that issues of hierarchy of command remained to be corrected in the years following the period under study.

Further refinement of multiship operations came with the cruise of the Squadron for Special Service in 1892–93 and the Naval Review Fleet in 1893. Both of these organizations provided extensive formation work and experience in the administrative and logistical problems associated with the deployment of large numbers of ships. However, while live-fire practice and drills were carried out on each ship, multiship formation work and maneuvers in both of these units were largely about appearance, not fighting skill. Appearance was import-ant in this age. The participation of the U.S. Navy on the stage of international naval pageantry made important contributions to the image of the United States in its people's eyes and abroad as well as to the squadron-based identity of U.S. Navy warships.

From 1895 on, squadron maneuvers focused more on combat. Maneuvers not only exercised the warships at various formations but also began to have a specific strategic purpose, such as the blockade of Charleston, South Carolina, in 1896. Target practice, however, continued to be at stationary targets. There is no evidence during this time that practice that married the ability to maneuver warships in close-order formation with accurate naval gunnery took place. The limitations of stationary target practice would be displayed at Santiago de Cuba. The outcome of that engagement would lead to gunnery reforms, most notably the work of Adm. William A. Sims, in the first decades of the twentieth century.

As the North Atlantic Squadron focused more on its mission to develop the capability to engage an enemy squadron at sea, it devoted more time and resources to the naval militia movement. Assigning the mission of the protec-tion of strategically important harbors and cities to ironclad monitors crewed by militia volunteers served two important purposes. Supporting local and state naval militias gave key local leaders, representatives, and senators along the East Coast a stake in the continued modernization of the Navy. It also freed the North Atlantic Squadron to carry out its primary mission, which was train-ing for the engagement and destruction of an enemy fleet at sea. During the Spanish-American-Cuban War, the presence of naval militia, although not as calming to the populace as the presence of a steel warship, enabled the forma-tion of the Northern Patrol Squadron, which allowed the North Atlantic and Flying squadrons to focus on taking the offense.

The unpacking of this process of organizational change in the North Atlantic Squadron has implications for the debate over the nature of U.S. imperial aspi-rations in the latter half of the nineteenth century. The results of this study rein-force and extend the work done by Stephen Roberts to claim that a "pattern of informal empire" existed in the years prior to the Spanish-American-Cuban War.[14] This study concludes that not only did the Navy regularly call at ports considered vital to U.S. business interests, as Roberts showed, but the Navy Department made a conscious decision as early as 1874 to develop the capabil-ity to fight fleet-level engagements. That capability was not fully attained during the period of this study, but the process of change provides evidence that the "imperial moment" of 1898 was not an accident. It was the result of a deliber-ate course of action and the development of a specific naval capability tied to the imposition of U.S. will beyond its borders. In this way, this study confirms the work of historians who have maintained that the events of 1898 can only be understood through the lens of the developments of the previous decades.[15]

Before either matériel or structural changes had taken place within the Navy Department, the North Atlantic Squadron was undergoing a process by which not only its function but also its very identity was changing. This was happening before it was evident to outsiders in the form of new ships. Today's U.S. Navy is in the midst of a period of profound change.[16] With the end of the Cold War and the rise of global terrorism, a Navy that was structurally and materially designed to fight the Soviet Union has been called upon instead to fight the War on Terror. What this means for the Navy of tomorrow remains to be seen. Within today's military, however, function and identity have been forced to change in response to new missions—missions often carried out using matériel designed for functions in keeping with an outdated strategic purpose. This study highlights the importance of considered strategic direction and allocation of resources. It shows that the process of organizational change can be well under way in advance of new matériel and cautions against the resulting conflict between actual and desired missions and functions. Above all, it demonstrates that any new capability not developed through rigorous exercise at sea may, in fact, not be a capability at all, but simply the appearance of one.

NOTES

Introduction

1. George W. Baer, *One Hundred Years of Sea Power: The U.S. Navy, 1890–1990* (Stanford, CA: Stanford University Press, 1994), 11.

2. Lance Buhl, "Maintaining an American Navy," in *In Peace and War: Interpretations of American Naval History,* ed. Kenneth J. Hagan, 30th Anniversary Edition (Westport, CT: Praeger Security International, 2008), 145–173.

3. Douglas Mark Haugen, *How Nineteenth-Century Naval Theorists Created America's Twentieth-Century Imperialist Policy* (Lewiston, NY: Edwin Mellen Press, 2010); Peter Karsten, *The Naval Aristocracy: The Golden Age of Annapolis and the Emergence of Modern American Navalism* (Annapolis: Naval Institute Press, 2008); Mark R. Shulman, *Navalism and the Emergence of American Sea Power, 1882–1893* (Annapolis: Naval Institute Press, 1995).

4. Core naval history: "Standard narrative histories of naval policy and operations . . . which establish the master plot." See Jon T. Sumida and David A. Rosenberg, "Machines, Men, Manufacturing, Management, and Money: The Study of Navies as Complex Organizations and the Transformation of Twentieth Century Naval History," in *Doing Naval History: Essays Towards Improvement,* ed. John B. Hattendorf, Naval War College Historical Monograph Series (Newport, RI: Naval War College Press, 1995). The term "Naval Revolution" belongs to Walter Herrick; see Walter Herrick, *The American Naval Revolution* (Baton Rouge: Louisiana State University Press, 1966).

5. Shulman, *Navalism,* 2.

6. An example is Ronald Spector, *Professors of War: The Naval War College and the Development of the Naval Profession,* ed. B. M. Simpson III, Naval War College Historical Monograph Series (Newport, RI: Naval War College Press, 1977).

7. Kenneth J. Hagan, *This People's Navy* (New York: Macmillan, 1991), 180–192; Herrick, *American Naval Revolution,* 30–85; Robert W. Love Jr., *History of the U.S. Navy,* vol. 1, *1775–1941* (Harrisburg, PA: Stackpole Books, 1992), 345–382; Alfred T. Mahan, *The Influence of Sea Power Upon History,* 12th ed. (Boston: Little, Brown, 1918); Donald W. Mitchell, *History of the Modern American Navy, from 1883 through Pearl Harbor* (New York: Alfred A. Knopf, 1946), 22–23; Harold Sprout and Margaret Sprout, *The Rise of American Naval Power, 1776–1918,* ed. Jack Sweetman, Classics of Naval Literature (Annapolis: Naval Institute Press, 1990), 214–257.

8. "Ancillary naval history consists of those studies that deal primarily with naval machines, men (including biography), manufacturing, and management." Sumida and Rosenberg, "Machines, Men."

9. Donald Chisholm, *Waiting for Dead Men's Shoes: Origins and Development of the U.S. Navy's Officer Personnel System, 1793–1941* (Stanford, CA: Stanford University Press, 2001); Karsten, *Naval Aristocracy.*

10. Frederick S. Harrod, *Manning the New Navy: The Development of a Modern Naval Enlisted Force, 1899–1940,* ed. Jon L. Wakelyn, Contributions in American History (Westport, CT: Greenwood Press, 1978).

11. William M. McBride, *Technological Change and the United States Navy, 1865–1945,* ed. Merritt Roe Smith, Johns Hopkins Studies in the History of Technology (Baltimore: Johns Hopkins University Press, 2000).

12. Baer comes the closest. In a single sentence concerning the designation of the Atlantic Fleet in 1907, he notes that "fleet formation was necessary. Maneuvers were conducted accordingly." What those maneuvers consisted of is left to the reader's imagination. See Baer, *One Hundred Years,* 24.

13. Exceptions to this rule are Kenneth J. Hagen, *American Gunboat Diplomacy and the Old Navy, 1877–1889,* Contributions in Military History (Westport, CT: Greenwood Press, 1973); Stephen S. Roberts, "An Indicator of Informal Empire: Patterns of U.S. Navy Cruising on Overseas Stations, 1869–1897," in *Fourth Naval History Symposium,* ed. Craig L. Symonds (Annapolis: Naval Institute Press, 1979).

14. Cf. Samuel Huntington, *The Soldier and the State: The Theory and Politics of Civil-Military Relations,* 12th ed. (Cambridge: Belknap Press of Harvard University Press, 1995), 228.

15. Sumida and Rosenberg, "Machines, Men," 31.

16. R. G. Collingwood and J. Dussen, *The Idea of History: With Lectures 1926–1928* (New York: Oxford University Press, 1994).

17. On this see, for example, Walter LaFeber, *The New Empire: An Interpretation of American Expansion, 1860–1898,* 4th ed. (Ithaca, NY: Cornell University Press, 1963), 61.

18. Milton Plesur, *America's Outward Thrust: Approaches to Foreign Affairs, 1865–1890* (DeKalb: Northern Illinois University Press, 1971).

19. It could be argued that the Pacific Squadron had a national defense role with respect to California and the Puget Sound region, but that hardly compared with the expectation that the North Atlantic Squadron would have the ability to protect cities vital to the national economy, such as New York.

20. Department of the Navy, *Annual Report of the Secretary of the Navy on the Operations of the Department, with Accompanying Documents for the Year 1866* (hereafter cited as Department of the Navy, *Annual Report of the Secretary of the Navy,* followed by the year).

21. A. P. Niblack, "Discussion of Prize Essay, 1895," U.S. Naval Institute *Proceedings* 21, no. 2 (1895).

22. James Phinney III Baxter, "June Meeting: The British High Commissioners at Washington 1871," *Proceedings of the Massachusetts Historical Society* 65 (October 1832–May 1936) and 3 (1936); J. Jay, *The Fisheries Dispute: A Suggestion for Its Adjustment by Abrogating the Convention of 1818, and Resting on the Rights and Liberties Defined in the Treaty of 1783; a Letter to the Honourable William M. Evarts, of the United States Senate* (New York: Dodd, Mead, 1887).

23. Louis A. Perez Jr., *The War of 1898: The United States and Cuba in History and Historiography* (Chapel Hill: University of North Carolina Press, 1998), 1–7.

24. Rayford W. Logan, *The Diplomatic Relations of the United States with Haiti, 1776–1891* (Chapel Hill: University of North Carolina Press, 1941); Ludwell Lee Montague, *Haiti and the United States, 1714–1938* (Durham, NC: Duke University Press, 1940).

25. Walter LaFeber, *The Panama Canal: The Crisis in Historical Perspective* (New York: Oxford University Press, 1989); John Lindsay-Poland, *Emperors in the Jungle: The Hidden History of the U.S. in Panama,* ed. Gilbert M. Joseph and Emily S. Rosenberg, American Encounters / Global Interactions series (Durham, NC: Duke University Press, 2003); Alfred Charles Richard Jr., *The Panama Canal in American National Consciousness, 1870–1990* (New York: Garland, 1990).

26. Rear Adm. Stephen B. Luce, "Journal of the Movements of the U.S.N.A. Fleet, Commencing February 3, 1874," Ship's Log, Papers of Stephen Bleeker Luce, Naval Historical Foundation Collection, Library of Congress, Washington, DC (hereafter cited as Luce Papers).

27. I am indebted to Tim Wolters for the concept of "cognitive experience of command." See Timothy Scott Wolters, "Managing a Sea of Information: Shipboard Command and Control in the United States Navy, 1899–1945" (Ph.D. diss., Massachusetts Institute of Technology, 2003).

28. Jan Rüger, *The Great Naval Game: Britain and Germany in the Age of Empire,* ed. Jay Winter, Studies in the Social and Cultural History of Modern Warfare (New York: Cambridge University Press, 2007).

Chapter 1. The North Atlantic Squadron, the *Virginius* Affair, and the Birth of Squadron Exercises, 1874–1881

1. The terms "naval forces on the North Atlantic Station" and "North Atlantic Squadron" are interchangeable and were used interchangeably by contemporaries. I will argue that the use of the term "squadron" increases in both popular print and official correspondence throughout the period under study, as the organization's identity evolves.

2. Foxhall A. Parker, "Our Fleet Maneuvers in the Bay of Florida, and the Navy of the Future," U.S. Naval Institute *Proceedings* 1, no. 8 (1874). I use the term "fleet" here, and throughout the study, as Commodore Parker did—as a way of describing tactical evolutions involving multiple warships. I do not mean to imply that simply by practicing these maneuvers the United States possessed a "battle fleet" at this time.

3. An exception to this is Love, *History of the U.S. Navy* 1:334–338. Professor Love argues that this episode (in 1874) highlights the fact that U.S. policy in the Caribbean could not move forward without a battle fleet.

4. "Line ahead": a group of ships following a guide in a straight line, one behind another. Contemporary U.S. tacticians referred to this formation in tactical guides as a "column," the term I will use throughout this study.

5. Wayne P. Hughes, *Fleet Tactics and Coastal Combat,* 2nd ed. (Annapolis: Naval Institute Press, 2000), 63–67.

6. For more on the impact of the torpedo on late nineteenth-century naval tactics, see Katherine C. Epstein, "No One Can Afford to Say 'Damn the Torpedoes': Battle Tactics and U.S. Naval History before World War I," *Journal of Military History* 77, no. 2 (April 2013): 491–520.

7. The following account is taken from Richard Hill, *War at Sea in the Ironclad Age,* ed. John Keegan, Cassell's History of Warfare series (London: Cassell, 2000). See also Spencer C. Tucker, *Handbook of 19th Century Naval Warfare* (Annapolis: Naval Institute Press, 2000), 131–133; Hughes, *Fleet Tactics and Coastal Combat,* 63–67.

8. The nomenclature designating different types of naval guns during this period can be confusing. In general, a gun primarily designed to fire shot (solid projectiles) was designated by the weight of its shot, for example, a "72-pounder." A gun primarily designed to fire shells (explosive projectiles) was designated in the U.S. Navy by the diameter of its bore, for example, "6 inch." These generalizations break down quickly, as many shell guns were capable of firing solid shot and vice versa. For more, see Tucker, *Handbook,* chap. 4.

9. It is interesting to note that Commo. Foxhall Parker, in addition to his more famous works on contemporary naval tactics, wrote a book on warfare in the age of galleys.

10. Sam Willis, "Fleet Performance and Capability in the Eighteenth-Century Royal Navy," *War in History* 11, no. 4 (2004): 373–92.

11. Hughes, *Fleet Tactics and Coastal Combat,* 64.

12. United States Naval Academy, "Annual Register of the U.S. Naval Academy, 1870," United States Naval Academy, Special Collections and Archives Department, Nimitz Library, Annapolis, MD.

13. The following is drawn from Clark G. Reynolds, *Famous American Admirals* (Annapolis: Naval Institute Press, 2002).

14. Parker died in 1879 while assigned as the superintendent of the U.S. Naval Academy. He is buried in the Naval Academy cemetery on Hospital Point—ironically, only feet from Rear Adm. Bancroft Gherardi's grave and a few yards from Capt. William Cushing's grave. One can only speculate about the heights Parker would have achieved in the New Steel Navy had he lived.

15. Foxhall A. Parker, *Squadron Tactics under Steam* (New York: D. Van Nostrand, 1864). In the special collection of Nimitz Library, United States Naval Academy.

16. Bureau of Navigation, circular order, 10 December 1869, Records of the Bureau of Naval Personnel, RG 24, NARA, Washington, DC; Foxhall A. Parker, *Fleet Tactics under Steam* (New York: D. Van Nostrand, 1870).

17. Parker, *Fleet Tactics under Steam,* 10.

18. On the debate over these various formations, see Hughes, *Fleet Tactics and Coastal Combat,* 70–75.

19. Parker, *Fleet Tactics under Steam,* 12–13, 217–218.

20. Namely, blockade of Southern ports and riverine operations in support of Union ground troops.

21. Farragut at Mobile Bay (1864) being the prime example of this. For a discussion of important naval actions during the U.S. Civil War and the argument that it was an important but unique naval conflict, see Hill, *War at Sea.*

22. All figures taken from Department of the Navy, *Annual Report of the Secretary of the Navy,* 1874.

23. Ibid.

24. Alfred T. Mahan, *From Sail to Steam: Recollections of Naval Life* (New York: Harper and Brothers, 1907), 271.

25. Ships with multiple cases of yellow fever on board were typically ordered to the Naval Hospital at Portsmouth, New Hampshire. See, for example, Secretary of the Navy George M. Robeson to Mullany, 18 July 1874, Letters to Flag Officers Commanding Squadrons, Vessels, and Stations, Sent by the Secretary of the Navy, 1867–1886, RG 45, NARA, Washington, DC.

26. Eugene Peltier, *The Bureau of Yards and Docks of the Navy and the Civil Engineer Corps* (New York: Newcomen Society in America, 1961); R. J. Winklareth, *Naval Shipbuilders of the World: From the Age of Sail to the Present Day* (London: Chatham, 2000), 125–209.

27. CinC J. C. Howell, Bureau of Ordnance, to Mullany, 18 October 1875, Letters to Flag Officers Commanding Squadrons, Vessels, and Stations, Sent by the Secretary of the Navy, 1867–1886, RG 45, NARA, Washington, DC. This letter is one example of the tensions between operational and yard commanders. There are many like it in this era.

28. Secretary of the Navy Adolph Borie to Dahlgren, 2 June 1869, U.S. Navy Department Letters Sent by the Secretary of the Navy to the Chiefs of Navy Bureaus, 1842–1886, RG 45, NARA, Washington, DC.

29. Adm. David Dixon Porter to Robeson, 4 December 1873, Records of the Bureau of Ordnance, RG 74, NARA, Washington, DC.

30. Tucker, *Handbook*, 168.

31. Adm. David Dixon Porter to Case, 12 June 1872, Records of the Bureau of Ordnance, RG 74, NARA, Washington, DC.

32. "Bigbadbattleships: Pre Dreadnaught Homeport," http://bigbadbattleships.com/ (accessed 10 February 2011).

33. Ibid.

34. European navies had not yet integrated torpedo boats into "grand fleet" tactics, either.

35. Lt. E. W. Very, USN, *Report on Torpedo-Boats for Coast Defense* (Washington, DC: GPO, 1884).

36. Rear Adm. James Alden to Case, 27 April 1872, Records of the Bureau of Ordnance, RG 74, NARA, Washington, DC; "Torpedoes and Sunken Mines," *Colburn's United Service Magazine* 1, no. 551 (1874).

37. E. O. Matthews to Alden, 15 April 1870, Records of the Bureau of Naval Personnel, RG 24, NARA, Washington, DC.

38. Bureau of Navigation, "General Order, Bureau of Navigation, 19 July 1869," Records of the Bureau of Naval Personnel, RG 24, NARA, Washington, DC.

39. Rear Adm. Daniel Ammen to Parker, 5 May 1874, Records of the Bureau of Naval Personnel, RG 24, NARA, Washington, DC; Rear Adm. Daniel Ammen to Parker, 15 October 1873, Records of the Bureau of Naval Personnel, RG 24, NARA, Washington, DC.

40. Commodore Worden, superintendent of the Naval Academy, asked for copies of the new signal manual for the purpose of teaching it to the students at the academy. See Commo. John J. Almy to Commo. John L. Worden, 1872, Records of the Bureau of Naval Personnel, RG 24, NARA, Washington, DC.

41. Commo. John J. Almy to Ammen, April 1872, Records of the Bureau of Naval Personnel, RG 24, NARA, Washington, DC.

42. The office of the chief signal officer arranged the rental of an ambulance, mules, and drivers to ferry naval officers receiving the training to Fort Whipple, where the U.S. Army's signal school was located. See Rear Adm. Daniel Ammen to Almy, 26 April 1872, Records of the Bureau of Naval Personnel, RG 24, NARA, Washington, DC.

43. Rear Adm. James Alden to Almy, 3 February 1871, Records of the Bureau of Naval Personnel, RG 24, NARA, Washington, DC.

44. There are multiple reports like this. See, for example, Commo. John J. Almy to Alden, 26 June 1871, Records of the Bureau of Naval Personnel, RG 24, NARA, Washington, DC.

45. Commo. John J. Almy to Luce, 4 May 1871, Records of the Bureau of Naval Personnel, RG 24, NARA, Washington, DC.

46. Commo. John J. Almy to Ammen, 31 October 1872, Records of the Bureau of Naval Personnel, RG 24, NARA, Washington, DC.

47. Commo. Foxhall Parker, "Omissions, Corrections, Additions to the U.S. Naval Signal Book (Steam Tactics) Proposed and Submitted by Capt. F. A. Parker, 1872," letter, Entry 78, Box 1, Records of the Bureau of Naval Personnel, RG 24, NARA, Washington, DC.

48. Commo. John J. Almy to Ammen, 10 January 1873, Records of the Bureau of Naval Personnel, RG 24, NARA, Washington, DC.

49. On Spain as a declining world power during this era, see Zbigniew Brzezinski, *Strategic Vision: America and the Crisis of Global Power* (New York: Basic Books, 2012), 16. As an example of the enormous financial toll that prolonged wars exacted from Spain, see Geoffrey Parker, *The Army of Flanders and the Spanish Road, 1567–1659,* Cambridge Studies in Early Modern History (New York: Cambridge University Press, 1972).

50. Capt. William Ronckendorff to Rear Admiral Scott, 3 July 1873, Records of the Operating Forces, RG 313, NARA, Washington, DC.

51. On the events leading up to the confrontation over the *Virginius,* see Richard H. Bradford, *The Virginius Affair* (Boulder: Colorado Associated University Press, 1980); Rear Adm. French Ensor Chadwick, *The Relations of the United States and Spain: Diplomacy* (New York: Charles Scribner's Sons, 1909).

52. Bradford, *Virginius Affair.*

53. "Aspinwall" was the name used by Europeans to refer to the city founded by Panama Rail Road president William Henry Aspinwall at the eastern terminus of the Panama Rail Road. Indigenous Panamanians called the city Colón to honor Christopher Columbus. Eventually the city was universally referred to as Colón. Throughout this study, I will conform to the contemporary usage of U.S. naval officers in their official reports.

54. Cdr. William Cushing to Scott, 7 October 1873, Records of the Operating Forces, RG 313, NARA, Washington, DC; Rear Adm. G. H. Scott to Robeson, 30 October 1873, Letters Received by the Secretary of the Navy from Commanding Officers, North Atlantic Squadron (1866–1885), RG 45, NARA, Washington, DC.

55. An officer rank no longer in use that was roughly the equivalent of today's lieutenant, junior grade.

56. William Cushing to Mary Edwards, May 20, 1860, quoted in Robert J. Schneller Jr., *Cushing: Civil War Seal,* ed. Dennis E. Showalter, Military Profiles series (Washington, DC: Brassey's, 2004).

57. Ralph J. Roske and Charles Van Doren, *Lincoln's Commando: The Biography of Cdr. William B. Cushing, USN,* Bluejacket Books (1957; reprint, Annapolis: Naval Institute Press, 1995).

58. "Further Particulars," *New York Times,* 3 November 1864.

59. Bradford, *Virginius Affair*, 49.

60. U.S. Attorney General George Henry Williams subsequently ruled that the *Virginius* was, in fact, the legal property of the Cuban rebels and, as such, had no right to be flying the U.S. flag. See Bradford, *Virginius Affair*.

61. Reprints of the letters between Cushing and the Spanish military officials in Bradford.

62. E. M. H. Edwards, *Commander William Barker Cushing of the United States Navy* (New York: F. Tennyson Neely, 1898).

63. Bradford tends more toward the latter; Edwards and Roske and Van Doren definitely present the former.

64. The authors of the best Cushing biography make this claim, without providing documentation. Without going through Spanish archival sources it is hard to substantiate whether or not the Spanish were familiar with Cushing. Cushing's attack on the *Albemarle* was sensationally covered in U.S. newspapers, and Cushing did receive a vote of thanks from Congress. It is not too much of a stretch to say that Spanish naval officers would have been aware of the mission. We have much better proof that U.S. officials were afraid of Cushing and his hotheadedness. See Roske and Van Doren, *Lincoln's Commando*, 293.

65. Cdr. D. L. Braine to Scott, 11 December 1873, Records of the Operating Forces, RG 313, NARA, Washington, DC.

66. Secretary of State Hamilton Fish, Diary, 14 November 1873, Papers of Hamilton Fish, Naval Historical Foundation Collection, Washington, DC.

67. Rear Adm. G. H. Scott to Robeson, 20 December 1873, Letters Received by the Secretary of the Navy from Commanders, North Atlantic Squadron (1866–1885), RG 45, NARA, Washington, DC.

68. Rear Adm. A. Ludlow Case to Robeson, 22 January 1874, Navy Department Letters Received by the Secretary of the Navy from Commanding Officers, North Atlantic Squadron (1866–1885), RG 45, NARA, Washington, DC.

69. Rear Adm. G. H. Scott to Secretary of the Navy Robeson, telegram, 1873, Roll 10, U.S. Navy Department Letters Received by the Secretary of the Navy from Commanding Officers, North Atlantic Squadron (1866–1885), August 1872–December 1873, RG 45, NARA, Washington, DC.

70. "Foreign Miscellany," *New York Times,* 25 August 1874.

71. A village near Salamanca in Spain. The ship's name commemorates a victory over French forces there in 1812, during the Peninsular War.

72. Roger Chesneau and Eugene M. Kolesnik, eds. *Conway's All the World's Fighting Ships, 1860–1905,* ed. Robert Gardiner (London: Conway Maritime Press, 1979), 381.

73. "The Spanish War Vessel Arapiles," *New York Times,* 25 November 1873.

74. "The Spanish Navy in War. Its Uselessness against Our Ports Attacks on Our Commerce," *New York Times,* 25 November 1873.

75. Rear Adm. A. Ludlow Case to Robeson, 3 January 1874, U.S. Navy Department Letters Received by the Secretary of the Navy from Commanders, North Atlantic Squadron, 1866–1885, RG 45, NARA, Washington, DC.

76. W. B. Cogar, *Dictionary of Admirals of the U.S. Navy: 1862–1900* (Annapolis: Naval Institute Press, 1989), 26; Reynolds, *Famous American Admirals,* 63.

77. Rear Adm. A. Ludlow Case to Jeffers, 23 November 1873, Records of the Bureau of Ordnance, RG 74, NARA, Washington, DC. Case tells his successor at the Bureau of Ordnance that he needs various weapons delivered to Key West immediately, including "torpedoes, wires, fuses, and the newest and best working Electric apparatus."

78. Case to Robeson, 22 January 1874.

79. Ibid.

80. Foxhall A. Parker, *The Naval Howitzer Ashore* (New York: D. Van Nostrand, 1866).

81. This would come back to haunt Rear Adm. J. R. M. Mullany when he asked the Navy Department to allow him to move his ships north for the sickly season later in 1874. The department refused permission, not wanting the warships too far from the Caribbean. See Robeson to Mullany, 18 July 1874.

82. Case to Robeson, 22 January 1874.

83. The following discussion is based on Luce, "Journal of the Movements of the U.S.N.A. Fleet."

84. Logbook, USS *Wyoming*, 22 December 1873–30 April 1874, Logs of U.S. Naval Ships, 1801–1915, Records of the Bureau of Naval Personnel, RG 24, NARA, Washington, DC.

85. The author has personally witnessed a near-collision between a guided-missile cruiser and an aircraft carrier, with collision alarms sounding on both ships, during formation work.

86. Logbook, *USS Wyoming*, entry for 6 February 1874.

87. Ibid., entry for 12 February 1874.

88. Ibid., entry for 15 February 1874.

89. Ibid., entry for 17 February 1874.

90. Ibid., entry for 20 February 1874.

91. Luce, "Journal of the Movements of the U.S.N.A. Fleet," entry for Wednesday, 24 February; Parker, "Our Fleet Maneuvers in the Bay of Florida," 169.

92. "I desire to call the attention of the officers . . . especially to that relating to the torpedo, which, in my judgment is to act so great a part in all future offensive and defensive operations. Although the Torpedo School at Newport has done much to bring the system to its present advanced condition, it is yet in its infancy, and I wish to invite a closer attention to it by all." *Army and Navy Journal,* 11 April 1874.

93. That is, a 90-degree turn. See Commo. Foxhall Parker to Case, 7 March 1874, Letters Received by the Secretary of the Navy from Commanders, North Atlantic Squadron (1866–1885), RG 45, NARA, Washington, DC.

94. *Army and Navy Journal,* 11 April 1874, 549.

95. Ibid.

96. "The Naval Review," *Army and Navy Journal,* 7 March 1874.

97. Ibid. See also Parker, "Our Fleet Maneuvers in the Bay of Florida."

98. Secretary of the Navy George M. Robeson to Case, March 1874, Letters to Flag Officers Commanding Squadrons, Vessels, and Stations, Sent by the Secretary of the Navy, 1867–1886, RG 45, NARA, Washington, DC; Secretary of the Navy George M. Robeson to Case, telegram, 7 April 1874, Letters to Flag Officers Commanding Squadrons, Vessels, and Stations, Sent by the Secretary of the Navy, 1867–1886, RG 45, NARA, Washington, DC.

99. Secretary of the Navy George M. Robeson to Mullany, 19 June 1874, Letters to Flag Officers Commanding Squadrons, Vessels, and Stations, Sent by the Secretary of the Navy, 1867–1886, RG 45, NARA, Washington, DC.

100. Ibid.

101. The files of the Bureau of Steam Engineering are full of letters from engineering officers requesting transfer to one of the larger, more comfortable cruising ships.

102. The only exception being a small landing exercise in October 1876.

103. Secretary of the Navy George M. Robeson to LeRoy, order, 28 January 1876, Letters to Flag Officers Commanding Squadrons, Vessels, and Stations, Sent by the Secretary of the Navy, 1867–1886, RG 45, NARA, Washington, DC.

104. In 1866 the paymaster at Port Royal expended over $136,000 of provisions, small stores, clothing, and other miscellaneous supplies. Paymaster Robert W. Allen, U.S. Navy, "Registers of Receipts and Disbursements for Provisions, Clothing, Small Stores, and Contingent Expenses by the Paymaster at U.S. Naval Depot Port Royal, SC, 1865–1874," Ledger, Records Collection of the Office of Naval Records and Library, Fiscal Records, 1798–1890, RG 45, NARA, Washington, DC.

105. The last day of active operations was apparently 24 July 1866; ibid. See also Secretary of the Navy Gideon Welles to Bureau Chiefs, 2 February 1867, U.S. Navy Department Letters Sent by the Secretary of the Navy to the Chiefs of Navy Bureaus, 1842–1886, RG 45, NARA, Washington, DC.

106. "When our vessels are driven from Key West by yellow fever, Port Royal is the nearest and safest harbor of refuge." Department of the Navy, *Annual Report of the Secretary of the Navy,* 1875.

107. Secretary of the Navy George M. Robeson to Mullany, 11 February 1876, Letters to Flag Officers Commanding Squadrons, Vessels, and Stations, Sent by the Secretary of the Navy, 1867–1886, RG 45, NARA, Washington, DC.

108. "Notes from Washington," *New York Times,* 3 September 1876, 2. See also Rear Adm. Daniel Ammen to Eastman, 21 April 1876, Office of Naval Records and Library, 1775–1910, RG 45, NARA, Washington, DC.

109. Ibid. This letter is endorsed by Captain English as the "senior officer present" at "naval depot" Port Royal. A year later, letters are endorsed by the "commanding officer, naval station Port Royal." For example, see the endorsement of Cdr. Robert R. Lewis to Eastman, 28 May 1877, Office of Naval Records and Library, 1775–1910, RG 45, NARA, Washington, DC.

110. Robeson to LeRoy, 20 January 1876.

111. Rear Adm. William E. LeRoy, General Order No. 6, 1876, Luce Papers.

112. Secretary of the Navy George M. Robeson to LeRoy, telegram, 14 March 1876, Letters to Flag Officers Commanding Squadrons, Vessels, and Stations, Sent by the Secretary of the Navy, 1867–1886, RG 45, NARA, Washington, DC.

113. CinC J. C. Howell, Bureau of Ordnance, to LeRoy, 16 March 1876, Letters to Flag Officers Commanding Squadrons, Vessels, and Stations, Sent by the Secretary of the Navy, 1867–1886, RG 45, NARA, Washington, DC.

114. Secretary of the Navy George M. Robeson to LeRoy, 1 April 1876, Letters to Flag Officers Commanding Squadrons, Vessels, and Stations, Sent by the Secretary of the Navy,

1867–1886, RG 45, NARA, Washington, DC. The department "suggested" the *Huron* for this duty.

115. Secretary of the Navy George M. Robeson to LeRoy, 24 June 1876, Letters to Flag Officers Commanding Squadrons, Vessels, and Stations, Sent by the Secretary of the Navy, 1867–1886, RG 45, NARA, Washington, DC; Secretary of the Navy George M. Robeson to LeRoy, telegram, 17 May 1876, Letters to Flag Officers Commanding Squadrons, Vessels, and Stations, Sent by the Secretary of the Navy, 1867–1886, RG 45, NARA, Washington, DC.

116. Rear Adm. Daniel Ammen to Trenchard, 2 September 1876, Letters to Flag Officers Commanding Squadrons, Vessels, and Stations, Sent by the Secretary of the Navy, 1867–1886, RG 45, NARA, Washington, DC; Secretary of the Navy George M. Robeson to LeRoy, 9 August 1876, Letters to Flag Officers Commanding Squadrons, Vessels, and Stations, Sent by the Secretary of the Navy, 1867–1886, RG 45, NARA, Washington, DC.

117. CinC J. C. Howell, Bureau of Ordnance, to Trenchard, telegram, 11 September 1876, Letters to Flag Officers Commanding Squadrons, Vessels, and Stations, Sent by the Secretary of the Navy, 1867–1886, RG 45, NARA, Washington, DC.

118. Department of the Navy, *Annual Report of the Secretary of the Navy*, 1876, 36–57. See also Edgar Stanton Maclay, *Reminiscences of the Old Navy: From the Journal and Private Papers of Captain Edward Trenchard, and Rear-Admiral Stephen Decatur Trenchard* (New York: G. P. Putnam's, 1898), 345.

119. What we think of today as the Marines' amphibious landing capability wasn't really institutionalized until the early twentieth century. Prior to that, a ship's capability to field an infantry force consisted largely of sailors. See Jack Shulimson, *The Marine Corps Search for a Mission, 1880–1898,* ed. Theodore A. Wilson, Modern War Studies (Lawrence: University Press of Kansas, 1993).

120. Department of the Navy, *Annual Report of the Secretary of the Navy*, 1877.

121. Secretary of the Navy Thompson to Trenchard, 26 July 1877, Letters to Flag Officers Commanding Squadrons, Vessels, and Stations, Sent by the Secretary of the Navy, 1867–1886, Roll 3, RG 45, NARA, Washington, D.C.

122. Department of the Navy, *Annual Report of the Secretary of the Navy*, 1878.

123. Commo. R. W. Shufeldt to Howell, 12 September 1878, Letters to Flag Officers Commanding Squadrons, Vessels, and Stations, Sent by the Secretary of the Navy, 1867–1886, RG 45, NARA, Washington, DC. J. C. Howell is not to be confused with John Adam Howell, who invented an early operational form of the self-propelled torpedo.

124. Commo. R. W. Shufeldt to Howell, telegram, 3 October 1878, Letters to Flag Officers Commanding Squadrons, Vessels, and Stations, Sent by the Secretary of the Navy, 1867–1886, RG 45, NARA, Washington, DC.

125. Commo. R. W. Shufeldt to Howell, 15 October 1878, Letters to Flag Officers Commanding Squadrons, Vessels, and Stations, Sent by the Secretary of the Navy, 1867–1886, RG 45, NARA, Washington, DC.

126. Ibid.

127. Secretary of the Navy R. W. Thompson to Howell, 21 January 1879, Letters to Flag Officers Commanding Squadrons, Vessels, and Stations, Sent by the Secretary of the Navy, 1867–1886, RG 45, NARA, Washington, DC.

128. Department of the Navy, "Dictionary of American Naval Fighting Ships," Naval Historical Center, http://www.history.navy.mil/danfs/w11/wyman-i.htm/ (accessed 3 February 2011).

129. Secretary of the Navy R. W. Thompson to Wyman, 17 January 1879, Letters to Flag Officers Commanding Squadrons, Vessels, and Stations, Sent by the Secretary of the Navy, 1867–1886, RG 45, NARA, Washington, DC.

130. Secretary of the Navy R. W. Thompson to Hall, telegram, 25 February 1879, Letters to Flag Officers Commanding Squadrons, Vessels, and Stations, Sent by the Secretary of the Navy, 1867–1886, RG 45, NARA, Washington, DC.

131. Secretary of the Navy R. W. Thompson to Wyman, 7 April 1879, Letters to Flag Officers Commanding Squadrons, Vessels, and Stations, Sent by the Secretary of the Navy, 1867–1886, RG 45, NARA, Washington, DC. Admiral Wyman's address is "Norfolk, VA" in this letter.

132. Secretary of the Navy R. W. Thompson to Wyman, telegram, 17 June 1879, Letters to Flag Officers Commanding Squadrons, Vessels, and Stations, Sent by the Secretary of the Navy, 1867–1886, RG 45, NARA, Washington, DC.

133. For the relationship between the Navy and the Central American region during this time, see Hagen, *American Gunboat Diplomacy,* chap. 8, "Defining American Interest in the Isthmus."

134. Ibid., 151–152.

135. Secretary of the Navy R. W. Thompson to Wyman, telegram, 29 July 1879, Letters to Flag Officers Commanding Squadrons, Vessels, and Stations, Sent by the Secretary of the Navy, 1867–1886, RG 45, NARA, Washington, DC. Also Secretary of the Navy R. W. Thompson to Wyman, 2 August 1879, Letters to Flag Officers Commanding Squadrons, Vessels, and Stations, Sent by the Secretary of the Navy, 1867–1886, RG 45, NARA, Washington, DC.

136. This is one of Nathan Miller's chapter titles. Nathan Miller, *The U.S. Navy: A History,* 3rd ed. (Annapolis: Naval Institute Press, 1997), 143.

Chapter 2. Toward a New Identity, 1882–1888

1. So called because their detractors surmised that they couldn't possibly be "full-blooded" Republicans.

2. David Healy, *James G. Blaine and Latin America* (Columbia: University of Missouri Press, 2001), 54–99.

3. On naval officers as uncompromising advocates of increased naval power, see Karsten, *Naval Aristocracy,* 385–389.

4. "Quick enough, if good enough."

5. Charles Belknap, "The Naval Policy of the United States," U.S. Naval Institute *Proceedings* 6, no. 13 (April 1880): 380.

6. Ibid., 386.

7. Ibid., 372.

8. Lt. Seaton Schroeder, "The Type of (I) Armored Vessel, (II) Cruiser, Best Suited to the Present Needs of the United States," U.S. Naval Institute *Proceedings* 7, no. 1 (1881).

9. Very eventually left the Navy and went to work for an arms manufacturer in France. For an example of his not-politically-correct correspondence, see Lieutenant Very, USN, to Wells, 5 January 1880, Records of the Bureau of Naval Personnel, RG 24, NARA, Washington, DC. On the idea that naval officers wrote conservatively to curry congressional favor, see Karsten, *Naval Aristocracy,* 312.

10. Paolo Coletta, ed., *American Secretaries of the Navy*, vol. 1, *1775–1913* (Annapolis: U.S. Naval Institute Press, 1980), 389–393.

11. Department of the Navy, *Annual Report of the Secretary of the Navy*, 1882, 29.

12. "Such vessels [ironclads] are absolutely needed for the defense of the country in time of war; and if Congress be willing to at once appropriate the large sum necessary for their construction, thoroughly efficient vessels can be designed and built in this country." Ibid., 36.

13. Ibid., 27–81. See also Coletta, *American Secretaries of the Navy*, 391–392.

14. "In making this statement, we do not wish to be understood that an efficient navy can be composed of unarmored vessels ." U.S. House of Representatives, *Condition of the Navy: Letter from the Secretary of the Navy, in Response to a Resolution from the House of Representatives, Requesting the Views of the Minority of the Commission to Consider the Condition of the Navy*, 47th Cong., 1st sess., 1882. Executive Document No. 30.

15. Ibid.

16. He eventually died in St. Petersburg. See Coletta, *American Secretaries of the Navy*, 393.

17. Adm. J. G. Walker to Luce, 3 May 1884, Official Orders, Papers of John Grimes Walker, Naval Historical Foundation Collection, Library of Congress, Washington, DC (hereafter cited as Walker Papers).

18. Miller, *U.S. Navy*, 149. See also Department of the Navy, *Annual Report of the Secretary of the Navy*, 1882, vol. 1, pp. 154–155.

19. Department of the Navy, *Annual Report of the Secretary of the Navy*, 1883, vol. 1, pp. 8–9.

20. Rear Adm. George H. Cooper to Chandler, 1 May 1882, U.S. Navy Department Letters Received by the Secretary of the Navy from Commanding Officers, North Atlantic Squadron, 1866–1885, RG 45, NARA, Washington, DC.

21. The naval school at Philadelphia was the direct predecessor of the U.S. Naval Academy.

22. Cogar, *Dictionary of Admirals of the U.S. Navy*, 33–34; Cooper obituary, *New York Tribune*, 18 November 1891.

23. "Naval Review off Fortress Monroe, *New York Times*, 25 April 1882.

24. Donald L. Canney, *The Old Steam Navy: Frigates, Sloops, and Gunboats, 1815–1885*, 2 vols. (Annapolis: Naval Institute Press, 1990), vol. 1. Other descriptions of ships in this section are from Canney as well.

25. Walker had been ordered to the bureau in late 1881. See Adm. J. G. Walker to Porter, 31 August 1881, Papers of David Dixon Porter, Naval Historical Foundation Collection, Library of Congress, Washington, DC. We know that Porter was happy to have him there and felt that he would be a positive influence on the secretary of the navy. See Adm. David Dixon Porter to Luce, 15 August 1881, Luce Papers.

26. Rear Adm. George H. Cooper to Chandler, 4 May 1882, U.S. Navy Department Letters Received by the Secretary of the Navy from Commanding Officers, North Atlantic Squadron, 1866–1885, RG 45, NARA, Washington, DC.

27. See, for instance, Rear Adm. George H. Cooper to Chandler, 30 May 1882, U.S. Navy Department Letters Received by the Secretary of the Navy from Commanding Officers, North Atlantic Squadron, 1866–1885, RG 45, NARA, Washington, DC.

28. Rear Adm. George H. Cooper, circular order, Cooper to Commanding Officers, 8 May 1882, U.S. Navy Department Letters Received by the Secretary of the Navy from Com-

manding Officers, North Atlantic Squadron, 1866–1885, RG 45, NARA, Washington, DC; Rear Adm. George H. Cooper, circular order, Cooper to Commanding Officers, 9 May 1882, U.S. Navy Department Letters Received by the Secretary of the Navy from Commanding Officers, North Atlantic Squadron, 1866–1885, RG 45, NARA, Washington, DC.

29. Capt. Earl English to Cooper, 25 May 1882, Records of the Naval Operating Forces, RG 313, NARA, Washington, DC.

30. On this, see Andrew Gordon, *The Rules of the Game: Jutland and British Naval Command* (London: John Murray, 2005), 434–437. See chapter 1 of this study for illustrations of these formations.

31. Rear Adm. George H. Cooper to Chandler, 1 June 1882, U.S. Navy Department Letters Received by the Secretary of the Navy from Commanding Officers, North Atlantic Squadron, 1866–1885, RG 45, NARA, Washington, DC.

32. Cooper to Commanding Officers, 9 May 1882, U.S. Navy Department Letters Received by the Secretary of the Navy from Commanding Officers, North Atlantic Squadron, 1866–1885, RG 45, NARA, Washington, DC.

33. Cooper to Chandler, 1 June 1882.

34. This observation is based on a review of the Department of the Navy's *Annual Report of the Secretary of the Navy* for the years 1875–81, as well as an analysis of the movements of the squadron flagship, *Tennessee,* as recorded in her logbooks.

35. Rear Adm. George H. Cooper to Chandler, 2 June 1882, U.S. Navy Department Letters Received by the Secretary of the Navy from Commanding Officers, North Atlantic Squadron, 1866–1885, RG 45, NARA, Washington, DC.

36. Rear Adm. George H. Cooper, circular letter to Commanding Officers, 2 June 1882, U.S. Navy Department Letters Received by the Secretary of the Navy from Commanding Officers, North Atlantic Squadron, 1866–1885, RG 45, NARA, Washington, DC.

37. It is during this period that newspapers begin to regularly refer to the "annual winter cruise" and "annual summer exercises."

38. Department of the Navy, *Annual Report of the Secretary of the Navy,* 1882, 230.

39. Rear Adm. George H. Cooper to Chandler, 17 June 1882, U.S. Navy Department Letters Received by the Secretary of the Navy from Commanding Officers, North Atlantic Squadron, 1866–1885, RG 45, NARA, Washington, DC.

40. Rear Adm. George H. Cooper to Chandler, 30 June 1882, U.S. Navy Department Letters Received by the Secretary of the Navy from Commanding Officers, North Atlantic Squadron, 1866–1885, RG 45, NARA, Washington, DC.

41. Logbook, USS *Tennessee,* 11 June 1882–9 December 1882, entry for 18 June, Logs of U.S. Naval Ships, 1801–1915, Records of the Bureau of Naval Personnel, RG 24, NARA, Washington, DC.

42. Ibid.

43. Rear Adm. George H. Cooper to Chandler, 22 June 1882, U.S. Navy Department Letters Received by the Secretary of the Navy from Commanding Officers, North Atlantic Squadron, 1866–1885, RG 45, NARA, Washington, DC.

44. Cooper to Chandler, 30 June 1882.

45. Rear Adm. George H. Cooper to Chandler, 12 July 1882, U.S. Navy Department Letters Received by the Secretary of the Navy from Commanding Officers, North Atlantic Squadron, 1866–1885, RG 45, NARA, Washington, DC.

46. Rear Adm. George H. Cooper to Chandler, 22 July 1882, U.S. Navy Department Letters Received by the Secretary of the Navy from Commanding Officers, North Atlantic Squadron, 1866–1885, RG 45, NARA, Washington, DC; Rear Adm. George H. Cooper, "North Atlantic Station General Order No. 11, 1882," U.S. Navy Department Letters Received by the Secretary of the Navy from Commanding Officers, North Atlantic Squadron, 1866–1885, RG 45, NARA, Washington, DC.

47. Rear Adm. George H. Cooper to Chandler, 2 August 1882, U.S. Navy Department Letters Received by the Secretary of the Navy from Commanding Officers, North Atlantic Squadron, 1866–1885, RG 45, NARA, Washington, DC.

48. Ibid.

49. Ibid.

50. Ibid.

51. Rear Adm. George H. Cooper to Chandler, 8 August 1882, U.S. Navy Department Letters Received by the Secretary of the Navy from Commanding Officers, North Atlantic Squadron, 1866–1885, RG 45, NARA, Washington, DC.

52. Rear Adm. George H. Cooper to Chandler, 12 August 1882, U.S. Navy Department Letters Received by the Secretary of the Navy from Commanding Officers, North Atlantic Squadron, 1866–1885, RG 45, NARA, Washington, DC.

53. Department of the Navy, General Order No. 301, 1882, Logs of U.S. Naval Ships, 1801–1915, Records of the Bureau of Naval Personnel, RG 24, NARA, Washington, DC. This was the same Lieutenant Very who had submitted the 1881 U.S. Naval Institute prize essay.

54. Commodore Johnson, USN, to Walker, 30 June 1882, Records of the Bureau of Naval Personnel, RG 24, NARA, Washington, DC; Commodore Johnson, USN, to Walker, 6 February 1884, Records of the Bureau of Naval Personnel, RG 24, NARA, Washington, DC; Commodore Johnson, USN, to Walker, 12 April 1884, Records of the Bureau of Naval Personnel, RG 24, NARA, Washington, DC.

55. Rear Adm. George H. Cooper, North Atlantic Station General Order No. 14, 1882, U.S. Navy Department Letters Received by the Secretary of the Navy from Commanding Officers, North Atlantic Squadron, 1866–1885, RG 45, NARA, Washington, DC.

56. Rear Adm. George H. Cooper to Chandler, 27 November 1882, U.S. Navy Department Letters Received by the Secretary of the Navy from Commanding Officers, North Atlantic Squadron, 1866–1885, RG 45, NARA, Washington, DC.

57. Rear Adm. George H. Cooper to Chandler, 1 March 1883, U.S. Navy Department Letters Received by the Secretary of the Navy from Commanding Officers, North Atlantic Squadron, 1866–1885, RG 45, NARA, Washington, DC; Rear Adm. George H. Cooper to Chandler, 5 February 1883, U.S. Navy Department Letters Received by the Secretary of the Navy from Commanding Officers, North Atlantic Squadron, 1866–1885, RG 45, NARA, Washington, DC; Rear Adm. George H. Cooper to Chandler, 8 January 1883, U.S. Navy Department Letters Received by the Secretary of the Navy from Commanding Officers, North Atlantic Squadron, 1866–1885, RG 45, NARA, Washington, DC;

Rear Adm. George H. Cooper to Chandler, 22 January 1883, U.S. Navy Department Letters Received by the Secretary of the Navy from Commanding Officers, North Atlantic Squadron, 1866–1885, RG 45, NARA, Washington, DC.

58. Rear Adm. George H. Cooper to Chandler, 13 March 1883, U.S. Navy Department Letters Received by the Secretary of the Navy from Commanding Officers, North Atlantic Squadron, 1866–1885, RG 45, NARA, Washington, DC.

59. Rear Adm. George H. Cooper to Chandler, 1 April 1883, U.S. Navy Department Letters Received by the Secretary of the Navy from Commanding Officers, North Atlantic Squadron, 1866–1885, RG 45, NARA, Washington, DC.

60. Rear Adm. George H. Cooper to Chandler, 20 May 1883, U.S. Navy Department Letters Received by the Secretary of the Navy from Commanding Officers, North Atlantic Squadron, 1866–1885, RG 45, NARA, Washington, DC; Rear Adm. George H. Cooper, North Atlantic Station General Order No. 23, 1883, U.S. Navy Department Letters Received by the Secretary of the Navy from Commanding Officers, North Atlantic Squadron, 1866–1885, RG 45, NARA, Washington, DC.

61. Rear Adm. George H. Cooper to Wallace, 20 June 1883, U.S. Navy Department Letters Received by the Secretary of the Navy from Commanding Officers, North Atlantic Squadron, 1866–1885, RG 45, NARA, Washington, DC.

62. Rear Adm. George H. Cooper to Chandler, 10 November 1883, U.S. Navy Department Letters Received by the Secretary of the Navy from Commanding Officers, North Atlantic Squadron, 1866–1885, RG 45, NARA, Washington, DC.

63. Cooper to Chandler, 22 June 1882.

64. Rear Adm. George H. Cooper, "Report of Condition and Employment of Vessels of the North Atlantic Squadron for the Month Ending August 31, 1883," U.S. Navy Department Letters Received by the Secretary of the Navy from Commanding Officers, North Atlantic Squadron, 1866–1885, RG 45, NARA, Washington, DC.

65. Rear Adm. George H. Cooper to Chandler, 22 June 1883, U.S. Navy Department Letters Received by the Secretary of the Navy from Commanding Officers, North Atlantic Squadron, 1866–1885, RG 45, NARA, Washington, DC.

66. Rear Adm. George H. Cooper to Chandler, 17 August 1883, U.S. Navy Department Letters Received by the Secretary of the Navy from Commanding Officers, North Atlantic Squadron, 1866–1885, RG 45, NARA, Washington, DC; Rear Adm. George H. Cooper to Chandler, 22 November 1883, U.S. Navy Department Letters Received by the Secretary of the Navy from Commanding Officers, North Atlantic Squadron, 1866–1885, RG 45, NARA, Washington, DC; Rear Adm. George H. Cooper, North Atlantic Station General Order No. 30, 1883, U.S. Navy Department Letters Received by the Secretary of the Navy from Commanding Officers, North Atlantic Squadron, 1866–1885, RG 45, NARA, Washington, DC.

67. Rear Adm. George H. Cooper to Chandler, 11 December 1883, U.S. Navy Department Letters Received by the Secretary of the Navy from Commanding Officers, North Atlantic Squadron, 1866–1885, RG 45, NARA, Washington, DC.

68. Rear Adm. George H. Cooper to Chandler, 16 May 1883, U.S. Navy Department Letters Received by the Secretary of the Navy from Commanding Officers, North Atlantic Squadron, 1866–1885, RG 45, NARA, Washington, DC.

69. For example, Luce's more famous subordinate, Alfred Thayer Mahan, was widely considered to be a subpar shiphandler and operational naval officer.

70. Luce was on duty at the Boston Navy Yard at the time, although he had briefly been detailed to command the *Minnesota* when it was thought that she would be put in commission during the *Virginius* crisis.

71. *Hartford,* along with *Plymouth, Vandalia, Marion,* and *Huron,* put ashore a combined 516 men and 6 guns. See Department of the Navy, *Annual Report of the Secretary of the Navy,* 1876, 36–37.

72. Spector, *Professors of War.*

73. John B. Hattendorf and John D. Hayes, eds., *The Writings of Stephen B. Luce,* Naval War College Historical Monograph Series (Newport, RI: Naval War College Press, 1975), 15.

74. Walker to Luce, 3 May 1884.

75. *Letter from the Secretary of the Navy Reporting, in Answer to Senate Resolution of the 4th Instant, the Steps Taken by Him to Establish an Advanced Course of Instruction of Naval Officers at Coasters' Harbor Island, Rhode Island,* 48th Cong., 2 sess., 1885.

76. Rear Adm. Stephen B. Luce to Chandler, 26 July 1884, U.S. Navy Department Letters Received by the Secretary of the Navy from Commanding Officers, North Atlantic Squadron, 1866–1885, RG 45, NARA, Washington, DC; Nichols to Luce, 22 July 1884, Offical Orders, Naval Historical Foundation Collection, Library of Congress, Washington, DC.

77. In 1881 U.S. Army lieutenant Adolphus W. Greely's polar expedition became stranded when their ship, *Proteus,* was stuck in the ice. After two failed attempts, a relief expedition, led by then-captain Winfield Scott Schley, rescued them on 20 June 1884 and returned them to the Naval Hospital at Portsmouth, New Hampshire. Of an original party of twenty-five men, six were rescued and five survived.

78. Hattendorf and Hayes, *Writings of Stephen B. Luce,* 195.

79. Rear Adm. Stephen B. Luce to Chandler, 18 August 1884, U.S. Navy Department Letters Received by the Secretary of the Navy from Commanding Officers, North Atlantic Squadron, 1866–1885, RG 45, NARA, Washington, DC.

80. Rear Adm. Stephen B. Luce to Chandler, 20 August 1884, U.S. Navy Department Letters Received by the Secretary of the Navy from Commanding Officers, North Atlantic Squadron, 1866–1885, RG 45, NARA, Washington, DC.

81. Ships' logs show the frustrations associated with this. For example, from the *Tennessee's* log, "To Alliance (Army): Watch our speed ball" (a speedball was a day shape indicating the flagship's speed), "To Fleet: Take position faster," and "To Vandalia: Get a better speed ball in next port." Logbook, USS *Tennessee,* entry for 21 June 1882.

82. See, for instance, Rear Adm. Stephen B. Luce to Walker, 29 August 1884, U.S. Navy Department Letters Received by the Secretary of the Navy from Commanding Officers, North Atlantic Squadron, 1866–1885, RG 45, NARA, Washington, DC.

83. "Army and Navy News," *New York Times,* 21 September 1884.

84. Cogar, *Dictionary of Admirals of the U.S. Navy,* 86–87; "Rear Admiral Jouett Dead," *New York Times,* 2 October 1902; Reynolds, *Famous American Admirals,* 170–171.

85. Rear Adm. James E. Jouett to Chandler, 4 March 1885, U.S. Navy Department Letters Received by the Secretary of the Navy from Commanding Officers, North Atlantic Squadron, 1866–1885, RG 45, NARA, Washington, DC.

86. On 7 March, William C. Whitney became the secretary of the navy in Democrat Grover Cleveland's first administration. Jouett corresponded with him immediately upon his return to *Tennessee*. Apparently some questions had been raised with the new secretary about Jouett's seniority and why he had been appointed to the North Atlantic Squadron. See Rear Adm. James E. Jouett to Whitney, 28 March 1885, U.S. Navy Department Letters Received by the Secretary of the Navy from Commanding Officers, North Atlantic Squadron, 1866–1885, RG 45, NARA, Washington, DC.

87. "The Burned City," *New York Times,* 2 April 1885.

88. U.S. House of Representatives, Committee on Foreign Affairs, *The Story of Panama: Hearings on the Rainey Resolution before the Committee on Foreign Affairs of the House of Representatives* (Washington, DC: GPO, 1913); U.S. Department of State, *Papers Relating to the Foreign Relations of the United States,* 1885, 244–245.

89. "Barrios Reported Dead," *New York Times,* 5 April 1885; Rear Adm. James E. Jouett to Whitney, 2 April 1885, U.S. Navy Department Letters Received by the Secretary of the Navy from Commanding Officers, North Atlantic Squadron, 1866–1885, RG 45, NARA, Washington, DC.

90. Jouett to Whitney, 2 April 1885.

91. Ibid.

92. Rear Adm. James E. Jouett to Vlloa, 10 April 1885, U.S. Navy Department Letters Received by the Secretary of the Navy from Commanding Officers, North Atlantic Squadron, 1866–1885, RG 45, NARA, Washington, DC.

93. "Notes from Central America," *New York Times,* 14 April 1885.

94. "The Panama Rebellion," *New York Times,* 4 April 1885.

95. Rear Adm. James E. Jouett to Whitney, 30 April 1885, U.S. Navy Department Letters Received by the Secretary of the Navy from Commanding Officers, North Atlantic Squadron, 1866–1885, RG 45, NARA, Washington, DC.

96. Rear Adm. James E. Jouett to Whitney, 24 April 1885, U.S. Navy Department Letters Received by the Secretary of the Navy from Commanding Officers, North Atlantic Squadron, 1866–1885, RG 45, NARA, Washington, DC; "Quiet Restored in Panama," *New York Times,* 26 April 1885.

97. Jouett to Whitney, 30 April 1885.

98. Rafael Reyes to Jouett, 5 May 1885, U.S. Navy Department Letters Received by the Secretary of the Navy from Commanding Officers, North Atlantic Squadron, 1866–1885, RG 45, NARA, Washington, DC.

99. Rear Adm. James E. Jouett to Whitney, 7 May 1885, U.S. Navy Department Letters Received by the Secretary of the Navy from Commanding Officers, North Atlantic Squadron, 1866–1885, RG 45, NARA, Washington, DC.

100. "The Situation in Panama," *New York Times,* 9 May 1885.

101. Rear Adm. James E. Jouett to Whitney, 11 May 1885, U.S. Navy Department Letters Received by the Secretary of the Navy from Commanding Officers, North Atlantic Squadron, 1866–1885, RG 45, NARA, Washington, DC.

102. Rear Adm. James E. Jouett to Whitney, 13 June 1885, U.S. Navy Department Letters Received by the Secretary of the Navy from Commanding Officers, North Atlantic Squadron, 1866–1885, RG 45, NARA, Washington, DC; "They Must Fight It Out," *New York Times,* 26 June 1885.

103. Rear Adm. James E. Jouett to Wildes, 14 July 1885, U.S. Navy Department Letters Received by the Secretary of the Navy from Commanding Officers, North Atlantic Squadron, 1866–1885, RG 45, NARA, Washington, DC.

104. "National Capital Topics," *New York Times,* 30 July 1885.

105. "Army and Navy News," *New York Times,* 24 July 1885.

106. Adm. J. G. Walker to McCalla, 1885, Walker Papers.

107. Jouett to Whitney, 7 May 1885.

108. Rear Adm. James E. Jouett to Whitney, 1 June 1885, U.S. Navy Department Letters Received by the Secretary of the Navy from Commanding Officers, North Atlantic Squadron, 1866–1885, RG 45, NARA, Washington, DC.

109. The secretary of the navy eventually decided that whoever the senior officer present (SOPA) was, regardless of side of the isthmus, would have tactical control over any navy units there, regardless of squadron. See Rear Adm. James E. Jouett, "North Atlantic Station General Order No. 31, 1885," U.S. Navy Department Letters Received by the Secretary of the Navy from Commanding Officers, North Atlantic Squadron, 1866–1885, RG 45, NARA, Washington, DC.

110. George C. Herring, *From Colony to Superpower: U.S. Foreign Relations since 1776,* ed. David M. Kennedy, Oxford History of the United States series (New York: Oxford University Press, 2008), 290–293.

111. Rear Adm. James E. Jouett to Whitney, 23 July 1885, U.S. Navy Department Letters Received by the Secretary of the Navy from Commanding Officers, North Atlantic Squadron, 1866–1885, RG 45, NARA, Washington, DC.

112. Secretary of the Navy Whitney to Jouett, 24 July 1885, U.S. Navy Department Letters Received by the Secretary of the Navy from Commanding Officers, North Atlantic Squadron, 1866–1885, RG 45, NARA, Washington, DC.

113. Rear Adm. James E. Jouett, North Atlantic Station General Order No. 29, 1885, U.S. Navy Department Letters Received by the Secretary of the Navy from Commanding Officers, North Atlantic Squadron, 1866–1885, RG 45, NARA, Washington, DC; "A Nation at a Tomb," *New York Times,* 9 August 1885.

114. Rear Adm. James E. Jouett to Whitney, 3 September 1885, U.S. Navy Department Letters Received by the Secretary of the Navy from Commanding Officers, North Atlantic Squadron, 1866–1885, RG 45, NARA, Washington, DC.

115. Rear Adm. James E. Jouett to Whitney, 10 November 1885, U.S. Navy Department Letters Received by the Secretary of the Navy from Commanding Officers, North Atlantic Squadron, 1866–1885, RG 45, NARA, Washington, DC.

116. Rear Adm. James E. Jouett to Whitney, 4 November 1885, U.S. Navy Department Letters Received by the Secretary of the Navy from Commanding Officers, North Atlantic Squadron, 1866–1885, RG 45, NARA, Washington, DC.

117. Adm. J. G. Walker to Jouett, 1 April 1886, Walker Papers.

118. Adm. David Dixon Porter to Luce, 21 March 1882, Luce Papers.

119. Captain Ramsey to Luce, 10 January 1884, Naval Historical Foundation Collection, Library of Congress, Washington, DC.

120. Hattendorf and Hayes, *Writings of Stephen B. Luce,* 13.

121. Capt. Alfred Thayer Mahan to Luce, 4 September 1884, Luce Papers. It is fitting that the building that houses NWC's operational studies and war-gaming facilities today bears McCarty Little's name.

122. Spector, *Professors of War*, 36.

123. Secretary of the Navy Whitney to Luce, 18 June 1886, Luce Papers.

124. Bowman Hendry McCalla to Luce, 7 December 1886, Luce Papers.

125. Capt. J. M. Ellicott, "Three Navy Cranks and What They Turned," U.S. Naval Institute *Proceedings* 50, no. 10 (1924): 1617.

126. Ibid.

127. Adm. J. G. Walker to Luce, 6 November 1886, Walker Papers.

128. In fact, the Navy Department was beginning to be concerned about hostilities with Great Britain, as evidenced by a curious exchange between Commodore Walker and another naval officer about doing reconnaissance on locks and canals to obstruct in Canada in the event of war. See Adm. J. G. Walker to Gridley, 25 July 1887, Walker Papers.

129. Secretary of the Navy Whitney to Luce, 20 June 1887, Luce Papers.

130. Edward H. Hart, *Squadron Evolutions: As Illustrated by the Combined Military and Naval Operations at Newport, R.I., November 1887, Rear Admiral S. B. Luce* (New York: E. H. Hart, 1887). See also Adm. J. G. Walker to Luce, 4 December 1887, Luce Papers.

131. Letter reprinted in Albert Gleaves, *Life and Letters of Rear Admiral Stephen B. Luce, U.S. Navy, Founder of the Naval War College* (New York: G. P. Putnam's Sons, 1925), 214.

132. Adm. J. G. Walker to Luce, 1 January 1888, Walker Papers.

133. Today's Dominican Republic.

134. Adm. J. G. Walker to Luce, 11 January 1888, reprinted in Gleaves, *Life and Letters,* 214.

135. He eventually moved onto *Galena,* a 1,900-ton steam sloop completely unsuited to carry an admiral's staff.

136. The contemporary correspondence all refers to the Republic of Haiti as "Hayti," the spelling common at the time. The modern spelling is used throughout the text, except in cases of direct quotation.

137. Rear Adm. Stephen B. Luce to Whitney, 28 July 1888, Luce Papers.

138. Ibid.

139. Rear Adm. Stephen B. Luce, "General Order, Naval Forces North Atlantic, 1888," Luce Papers.

140. Bayard to Whitney, 3 January 1889, letter reprinted in Gleaves, *Life and Letters,* 222. See also "Admiral Luce's Report," *New York Times*, 2 January 1889.

141. This is the title of an article by Robert Seager II. See "Ten Years Before Mahan: The Unofficial Case for the New Navy, 1880–1890," *Mississippi Valley Historical Review* 40, no. 3 (December 1953): 491–512.

142. Hattendorf and Hayes, *Writings of Stephen B. Luce,* 13.

Chapter 3. The North Atlantic Squadron and the Squadron of Evolution, 1889–1891

1. Frederick Jackson Turner, "The Significance of the Frontier in American History," in *History, Frontier, and Section: Three Essays by Frederick Jackson Turner,* by Frederick Jackson Turner (Albuquerque: University of New Mexico Press, 1993); Robert L. Beisner, *From the Old Diplomacy to the New, 1865–1900* (Arlington Heights, IL: Harlan Davidson, 1986); Herring, *From Colony to Superpower,* 290–295.

2. Secretary of the Navy Whitney to Luce, 28 January 1889, Luce Papers.

3. "Admiral Gherardi Transferred," *New York Times,* 10 February 1889, 5.

4. Department of the Navy, *Annual Report of the Secretary of the Navy,* 1889, vol. 1, p. 78.

5. Department of the Navy, *Annual Report of the Secretary of the Navy*, 1885, vol. 1, p. 207.

6. For strategic interest in Haiti, see Love, *History of the U.S. Navy* 1:363–365. On Galena's cruise, see "Back from Hayti," *New York Times*, 30 May 1889, 1.

7. "Ordered to Sea," *New York Times*, 14 June 1889, 8.

8. "Of Naval Interest," *New York Times*, 28 August 1889, 2. In seeming critical of Rear Admiral Gherardi, it must be remembered that the *Times* was a strong supporter of Luce and his various projects.

9. Rear Adm. Robley Evans, *A Sailor's Log: Recollections of Forty Years of Naval Life* (New York: D. Appleton,), 241. See also "Douglass's Trip to Hayti," *New York Times*, 30 September 1889, 5.

10. "Ready for Mr. Douglass," *New York Times*, 1 October 1889, 5.

11. "The Kearsarge Sails," *New York Times*, 2 October 1889, 8.

12. John D. Alden, *The American Steel Navy: A Photographic History of the U.S. Navy from the Introduction of the Steel Hull in 1883 to the Cruise of the Great White Fleet, 1907–1909* (Annapolis: Naval Institute Press, 1989), 16.

13. Rear Adm. Stephen B. Luce, "Our Future Navy," *North American Review* 149, no. 392 (1889): 65. See also the remarks by Sampson about coastal defense for a different viewpoint. "Capt. Sampson's Scheme," *New York Times*, 2 August 1889, 2.

14. There was a lot of chatter over this subject in the *New York Times* during the spring and summer of 1889. As examples, see "That Flying Squadron: Commodore Walker Favors a World Cruise," *New York Times*, 9 May 1889, 1; "Work at the Navy Yard: A Decidedly Mythical Cruise," *New York Tribune*, 12 May 1889, 3.

15. See chapter 2 of this volume.

16. "Editorial Article 5—No Title," *New York Times*, 20 July 1889, 4. Also "Naval Speed Premiums," *New York Times*, 8 August 1889, 4.

17. Commo. John G. Walker, by Cdr. C. F. Goodrich, to Walker, telegram, 21 September 1889, and Cdr. C. F. Goodrich to Walker, telegram, 24 September 1889, Area Files of the Naval Records Collection, 1775–1910, 1775–1910, RG 45, NARA, Washington, DC.

18. Daniel Howard Wicks, "New Navy and New Empire: The Life and Times of John Grimes Walker" (Ph.D. diss., University of California, Berkeley, 1979).

19. Adm. J. G. Walker to SECNAV, 17 September 1889, in Department of the Navy, Situation Reports, Squadron of Evolution, Dec 7, 1889–May 25, 1892, Records Collection of the Office of Naval Records and Library, RG 45, NARA, Washington, DC.

20. Rear Adm. George H. Cooper, "Signal Instructions, 1882," U.S. Navy Department Letters Received by the Secretary of the Navy from Commanding Officers, North Atlantic Squadron, 1866–1885, RG 45, NARA, Washington, DC.

21. "Practical Naval Work," *New York Times*, 6 October 1889, 16.

22. "Army and Navy News," *New York Times*, 2 April 1889, 2; "That Flying Squadron," *New York Times*, 9 May 1889, 1.

23. "Admiral Walker's Departure," *Washington Post*, 1 November 1889, 5.

24. "A New Cruiser Disabled," *New York Times*, 4 August 1889, 1.

25. "The Squadron Sails," *New York Times*, 8 December 1889, 5.

26. Adm. J. G. Walker to SECNAV, telegram, 18 November 1889, Area Files of the Naval Records Collection, 1775–1910, 1775–1910, RG 45, NARA, Washington, DC.

27. Department of the Navy, Situation Reports, Squadron of Evolution, Dec 7, 1889–May 25, 1892.

28. Adm. J. G. Walker to SECNAV, telegram, 7 December 1889, Area Files of the Naval Records Collection, 1775–1910, 1775–1910, RG 45, NARA, Washington, DC.

29. Wicks, "New Navy and New Empire," 218.

30. "War Ships to Be Proud Of," *New York Times,* 13 October 1889, 13.

31. Adm. J. G. Walker to SECNAV, telegram, 1889, Area Files of the Naval Records Collection, 1775–1910, 1775–1910, RG 45, NARA, Washington, DC.

32. Adm. J. G. Walker to SECNAV, 21 December 1889, Area Files of the Naval Records Collection, 1775–1910, 1775–1910, RG 45, NARA, Washington, DC.

33. Doris D. Maguire, ed., *French Ensor Chadwick: Selected Letters and Papers* (Washington, DC: University Press of America, 1981). See especially 16–143.

34. French Ensor Chadwick, *Report on the Training Systems for the Navy and Mercantile Marine of England and on the Naval Training System of France Made to the Bureau of Equipment and Recruiting, U.S. Navy Department, September 1879,* 46th Cong., 2nd sess., 1880.

35. Adm. J. G. Walker to SECNAV, 23 December 1889, Area Files of the Naval Records Collection, 1775–1910, 1775–1910, RG 45, NARA, Washington, DC.

36. This is the same Commander McCalla who had been Jouett's ground forces commander in Panama in 1885. Apparently his disagreements with Jouett, covered in chapter 2, did not affect his ability to gain command of a ship. See Bowman Hendry McCalla to SECNAV, 29 December 1889, Area Files of the Naval Records Collection, 1775–1910, 1775–1910, RG 45, NARA, Washington, DC.

37. Adm. J. G. Walker to SECNAV, 31 December 1889, Area Files of the Naval Records Collection, 1775–1910, 1775–1910, RG 45, NARA, Washington, DC.

38. Adm. J. G. Walker to SECNAV, 2 January 1890, Area Files of the Naval Records Collection, 1775–1910, 1775–1910, RG 45, NARA, Washington, DC.

39. Ibid.

40. Rear Adm. Bancroft Gherardi to Tracy, 29 December 1889, Area Files of the Naval Records Collection, 1775–1910, 1775–1910, RG 45, NARA, Washington, DC.

41. Alden, *American Steel Navy,* 14.

42. For all the trouble associated with building *Dolphin* and getting the Navy Department to accept her, see Leonard Alexander Swann Jr., *John Roach Maritime Entrepreneur: The Years as Naval Contractor, 1862–1886* (Annapolis: Naval Institute Press, 1965).

43. Fredrick Douglass to Blaine, 11 February 1890, Area Files of the Naval Records Collection, 1775–1910, 1775–1910, RG 45, NARA, Washington, DC.

44. Cdr. G. W. Sumner to Tracy, 17 February 1890, Area Files of the Naval Records Collection, 1775–1910, 1775–1910, RG 45, NARA, Washington, DC.

45. Rear Adm. Bancroft Gherardi to Sumner, 13 February 1890, Area Files of the Naval Records Collection, 1775–1910, 1775–1910, RG 45, NARA, Washington, DC.

46. Cdr. G. W. Sumner to Bureau of Navigation and Detail, telegram, 5 March 1890, Area Files of the Naval Records Collection, 1775–1910, 1775–1910, RG 45, NARA, Washington, DC; Cdr. G. W. Sumner to Bureau of Navigation and Detail, telegram, 7 March 1890, Area Files of the Naval Records Collection, 1775–1910, 1775–1910, RG 45, NARA, Washington, DC.

47. Department of the Navy, *Annual Report of the Secretary of the Navy,* 1890, 165–167. See also Cdr. G. H. Rockwell to Tracy, 8 March 1890, Area Files of the Naval Records Collection, 1775–1910, 1775–1910, RG 45, NARA, Washington, DC.

48. James G. Blaine to Tracy, 13 March 1890, Area Files of the Naval Records Collection, 1775–1910, 1775–1910, RG 45, NARA, Washington, DC.

49. Rear Adm. Bancroft Gherardi to Tracy, 24 March 1890, Area Files of the Naval Records Collection, 1775–1910, 1775–1910, RG 45, NARA, Washington, DC.

50. Cdr. G. W. Sumner to Tracy, 6 April 1890, Area Files of the Naval Records Collection, 1775–1910, 1775–1910, RG 45, NARA, Washington, DC.

51. Rear Adm. Bancroft Gherardi to Tracy, 18 April 1890, Area Files of the Naval Records Collection, 1775–1910, 1775–1910, RG 45, NARA, Washington, DC.

52. Rear Adm. Bancroft Gherardi to Tracy, 3 May 1890, Area Files of the Naval Records Collection, 1775–1910, 1775–1910, RG 45, NARA, Washington, DC.

53. Excluding *Dolphin,* which was not designed to be a "warship" in the best sense of the word.

54. "News of the Navies," *New York Times,* 11 April 1890, 3.

55. Gherardi to Tracy, 3 May 1890.

56. Fredrick Douglass to Blaine, 14 May 1890, Area Files of the Naval Records Collection, 1775–1910, 1775–1910, RG 45, NARA, Washington, DC.

57. Rear Adm. Bancroft Gherardi to Tracy, 29 May 1890, Area Files of the Naval Records Collection, 1775–1910, 1775–1910, RG 45, NARA, Washington, DC.

58. Contemporary reports concerning the results of target practice have to be taken with a grain of salt, as it was commonplace for umpires to record near-misses of the target as "hits," the logic being that an actual ship would be much larger than the target. Nonetheless, the fact remains that everyone who witnessed this exercise was pleased with the results. See B. A. Fiske, *From Midshipman to Rear-Admiral* (New York: Century, 1919), 123–124; "Our Squadron in Port," *New York Times,* 13 June 1890, 8.

59. Alden, *American Steel Navy,* 26.

60. Department of the Navy, *Annual Report of the Secretary of the Navy,* 1890.

61. "Of Naval Interest," *New York Times,* 17 October 1890, 9.

62. In keeping with the contemporary usage of naval officers in their correspondence, I (as they did) now switch reference to the city of Aspinwall to the Spanish name chosen by its mestizo inhabitants, Colón.

63. Cdr. Horace Elmer to Tracy, 16 September 1890, Area Files of the Naval Records Collection, 1775–1910, 1775–1910, RG 45, NARA, Washington, DC.

64. Cdr. Horace Elmer to Tracy, 22 September 1890, Area Files of the Naval Records Collection, 1775–1910, 1775–1910, RG 45, NARA, Washington, DC.

65. Cdr. Horace Elmer to Tracy, 4 October 1890, Area Files of the Naval Records Collection, 1775–1910, 1775–1910, RG 45, NARA, Washington, DC; Cdr. Horace Elmer to Tracy, 24 October 1890, Area Files of the Naval Records Collection, 1775–1910, 1775–1910, RG 45, NARA, Washington, DC.

66. William P. Clyde to Ramsey, 7 October 1890, Area Files of the Naval Records Collection, 1775–1910, 1775–1910, RG 45, NARA, Washington, DC.

67. Cdr. G. A. Converse to Tracy, 15 November 1890, Area Files of the Naval Records Collection, 1775–1910, 1775–1910, RG 45, NARA, Washington, DC; Rear Adm. Bancroft

Gherardi to Tracy, 12 December 1890, Area Files of the Naval Records Collection, 1775–1910, 1775–1910, RG 45, NARA, Washington, DC.

68. "To Winter at the Indies," *New York Times,* 12 December 1890, 9.

69. Ibid. See also "Of Naval Interest," *New York Times,* 12 December 1890, 5.

70. G. L. C., "Bluejackets on Shore," *New York Times,* 26 January 1890, 14.

71. Alberto Molina to Blaine, 24 January 1890, Area Files of the Naval Records Collection, 1775–1910, 1775–1910, RG 45, NARA, Washington, DC.

72. Adm. J. G. Walker to SECNAV, 3 March 1890, Department of the Navy, Situation Reports, Squadron of Evolution, Dec 7, 1889–May 25, 1892. See also "Article 1—No Title," *New York Times,* 2 March 1890, 2.

73. Adm. J. G. Walker to SECNAV, 16 March 1890, Area Files of the Naval Records Collection, 1775–1910, 1775–1910, RG 45, NARA, Washington, DC.

74. Adm. J. G. Walker to SECNAV, 29 March 1890, Area Files of the Naval Records Collection, 1775–1910, 1775–1910, RG 45, NARA, Washington, DC.

75. A. Loudon Snowden to Blaine, 18 April 1890, Area Files of the Naval Records Collection, 1775–1910, 1775–1910, RG 45, NARA, Washington, DC.

76. Adm. J. G. Walker to SECNAV, 28 April 1890, Area Files of the Naval Records Collection, 1775–1910, 1775–1910, RG 45, NARA, Washington, DC.

77. Wicks, "New Navy and New Empire," 238.

78. Adm. J. G. Walker to SECNAV, 17 May 1890, Area Files of the Naval Records Collection, 1775–1910, 1775–1910, RG 45, NARA, Washington, DC. See also "The Cruise Was a Success," *New York Times,* 26 May 1890, 5.

79. Adm. J. G. Walker to SECNAV, 30 May 1890, Area Files of the Naval Records Collection, 1775–1910, 1775–1910, RG 45, NARA, Washington, DC.

80. Adm. J. G. Walker to SECNAV, 6 June 1890, Area Files of the Naval Records Collection, 1775–1910, 1775–1910, RG 45, NARA, Washington, DC.

81. "Honors Due to Brazil: The Coming Visit of the New Republic's Squadron," *New York Times,* 12 November 1890, 9; "Squadron of Evolution: Received with Honors in the Ports of Brazil," *New York Times,* 24 July 1890, 3.

82. "What of the Squadron? Winter Maneuvers—Rear Admiral Walker's Position," *New York Times,* 20 September 1890, 8.

83. Rear Adm. Bancroft Gherardi to Ramsey, telegram, 3 January 1891, Area Files of the Naval Records Collection, 1775–1910, 1775–1910, RG 45, NARA, Washington, DC.

84. "Various Naval Items, 24 Jan 91," *Army and Navy Journal,* 24 January 1891.

85. Cdr. Horace Elmer to Gherardi, 13 January 1891, Area Files of the Naval Records Collection, 1775–1910, 1775–1910, RG 45, NARA, Washington, DC; Rear Adm. Bancroft Gherardi to Tracy, 5 January 1891, Area Files of the Naval Records Collection, 1775–1910, 1775–1910, RG 45, NARA, Washington, DC.

86. "The Navy: Naval Vessels in Commission, Where and When Last Heard From," *Army and Navy Journal,* 3 January 1891.

87. Rear Adm. Bancroft Gherardi to SECNAV, telegram, 15 January 1891, Area Files of the Naval Records Collection, 1775–1910, 1775–1910, RG 45, NARA, Washington, DC.

88. Rear Adm. Bancroft Gherardi to Tracy, 27 January 1891, Area Files of the Naval Records Collection, 1775–1910, 1775–1910, RG 45, NARA, Washington, DC.

89. Rear Adm. Bancroft Gherardi to Bureau of Navigation and Detail, telegram, 5 February 1891, Area Files of the Naval Records Collection, 1775–1910, 1775–1910, RG 45, NARA, Washington, DC.

90. Rear Adm. Bancroft Gherardi to Tracy, 26 February 1891, Area Files of the Naval Records Collection, 1775–1910, 1775–1910, RG 45, NARA, Washington, DC. See also Montague, *Haiti and the United States.*

91. "Various Naval Items, 18 Apr 91," *Army and Navy Journal,* 18 April 1891.

92. Adm. J. G. Walker to SECNAV, 13 April 1891, Area Files of the Naval Records Collection, 1775–1910, 1775–1910, RG 45, NARA, Washington, DC.

93. Rear Adm. Bancroft Gherardi to SECNAV, 18 April 1891, Area Files of the Naval Records Collection, 1775–1910, 1775–1910, RG 45, NARA, Washington, DC.

94. To steam into the presence of the flagship of a senior admiral (Gherardi was No. 2 on the list at this time) and not at the *very least* request permission to "proceed on duties assigned" from the senior officer would be considered an eyebrow-raising discourtesy even today, when these sorts of things are usually done over the Internet via instant chat or e-mail. Walker's actions must have been utterly shocking in an era of social rigidness and strict attention to etiquette.

95. "Gherardi's Stern Rebuke," *New York Times,* 23 August 1891, 1. For a first-person account from a junior officer who witnessed the event, see Edward L. Beach Sr. and Edward L. Beach Jr., *From Annapolis to Scapa Flow: The Autobiography of Edward L. Beach, Sr.* (Annapolis: Naval Institute Press, 2003), 59.

96. Department of the Navy, Situation Reports, Squadron of Evolution, Dec 7, 1889–May 25, 1892.

97. Beach and Beach, *From Annapolis to Scapa Flow,* 52.

98. Adm. J. G. Walker to Tracy, 29 April 1891, Area Files of the Naval Records Collection, 1775–1910, 1775–1910, RG 45, NARA, Washington, DC. See also "Various Naval Items, 9 May 91," *Army and Navy Journal,* 9 May 1891.

99. Montague, *Haiti and the United States,* 150–152.

100. Rear Adm. Bancroft Gherardi to Tracy, 3 May 1891, Area Files of the Naval Records Collection, 1775–1910, 1775–1910, RG 45, NARA, Washington, DC.

101. Probably either meningitis or encephalitis—both viral and neither caused by the climate. In any event, the *Army and Navy Journal* reported, "He was a man conscientious to the extreme in the matter of duty and might have come North before but manfully stuck to his post." A very typical Victorian-era linkage of gender to certain personality traits; see Gail Bederman, *Manliness and Civilization: A Cultural History of Gender and Race in the United States, 1880–1917,* Women in Culture and Society series (Chicago: University of Chicago Press, 1995).

102. "Recent Deaths, 23 May 91," *Army and Navy Journal,* 23 May 1891.

103. Adm. J. G. Walker to Cdr. J. W. Phillip, 11 March 1889, Walker Papers.

104. "Various Naval Items, 6 Jun 91," *Army and Navy Journal,* 6 June 1891.

105. "The Navy: Naval Vessels in Commission—When and Where Last Heard from, 4 Jul 91," *Army and Navy Journal,* 4 July 1891.

106. The others being the *Charleston, Baltimore, Philadelphia,* and *San Francisco.*
107. Alden, *American Steel Navy,* 25.
108. The state flag would not have been part of *Newark's* signaling equipment. Walker obviously sent a subordinate out in New York with orders to find and purchase a Massachusetts state flag just for this occasion, prior to departing for Boston. It says a lot about Walker, the kinds of things he focused on, and what he felt the role of his squadron was.
109. "The White Squadron's Attack on Fisher's Island," *Army and Navy Journal,* 8 August 1891.
110. "The Boston Naval Militia Exercises: Admiral Walker's Report to the Navy Dept.," *Army and Navy Journal,* 1 August 1891; "Honors to the Squadron of Evolution," *Army and Navy Journal,* 8 August 1891; Department of the Navy, Situation Reports, Squadron of Evolution, Dec 7, 1889–May 25, 1892; "The White Squadron at Boston," *Army and Navy Journal,* 18 July 1891.
111. Alden, *American Steel Navy,* 39.
112. "Admiral Walker's 'Pull,'" *New York Times,* 24 August 1891, 1; "Reducing Walker's Fleet," *New York Times,* 11 September 1891, 1; "Tracy Bows to the Pull," *New York Times,* 5 September 1891, 5; "Walker's Hurtful Pull," *New York Times,* 3 September 1891, 5; "Walker's Singular Pull," *New York Times,* 4 September 1891, 4; "Walker Still Defiant," *New York Times,* 26 August 1891, 1.
113. "The Navy: Naval Vessels in Commission, When and Where Last Heard from, 15 Aug 91," *Army and Navy Journal,* 15 August 1891.
114. "The Navy: Naval Vessels in Commission, When and Where Last Heard from, 29 Aug 91," *Army and Navy Journal,* 29 August 1891.
115. "Various Naval Items, 31 Oct 91," *Army and Navy Journal,* 31 October 1891.
116. Even in the twenty-first century, naval leadership is cautious about where they send sailors ashore to blow off steam—particularly when they have been at sea for a period of several months with no liberty. A port in a country that we had a few weeks earlier almost gone to a kinetic confrontation with would probably not be considered. In Schley's defense, his men were getting restless, and he was trying to maintain order on his ship, but here he showed the same sort of poor decision making that would cause him much grief at Santiago Bay seven years later. See Love, *History of the U.S. Navy* 1:367.
117. Secretary of State Blaine, who probably would have been a calming influence on Harrison, was out of town at the time.
118. F. Pratt, *Preble's Boys: Commodore Preble and the Birth of American Sea Power* (New York: Sloane, 1950).
119. For a good discussion of the problems, issues, and personal animosities associated with multiple senior officers and squadrons on the same station in the early Navy, see Stephen Budiansky, *Perilous Fight: America's Intrepid War with Britain on the High Seas, 1812–1815* (New York: Alfred A. Knopf, 2010). Maybe the best-known result of this mindset in the old navy is the death of Commo. Stephen Decatur in 1820 in a duel with Commo. James Barron.
120. "Wanted: A Naval Policy," *Army and Navy Journal,* 5 December 1891.
121. "The Value of Naval Maneuvers," *Army and Navy Journal,* 15 August 1891.

Chapter 4. The Limits of Ad Hoc Crisis Response, 1892–1894

1. Rüger, *Great Naval Game*, 1.
2. On the use of a navy to build national identity, see Jan Rüger, "Nation, Empire, and Navy: Identity Politics in the United Kingdom, 1887–1914," *Past and Present* 185 (2004): 159–188. For different interpretations of the development of post–Civil War U.S. national identity, see D. W. Blight, *Race and Reunion: The Civil War in American Memory* (Cambridge: Belknap Press of Harvard University Press, 2001); D. G. Faust, *This Republic of Suffering: Death and the American Civil War* (New York: Vintage Books, 2009).
3. *New York Advertiser*, 19 January 1892, as quoted in B. F. Cooling, *Benjamin Franklin Tracy: Father of the Modern American Fighting Navy* (Hamden, CT: Archon Books, 1973), 119.
4. "Take no part in troubles further than to protect American interests . . . use every precaution to avoid if possible such measures." Benjamin F. Tracy to McCann, telegram, 4 March 1891, Naval Records Collection of the Office of Naval Records and Library, 1691–1945, RG 45, NARA, Washington, DC.
5. Cooling, *Benjamin Franklin Tracy*, 121.
6. Benjamin F. Tracy to Walker, telegram, 10 December 1891, Naval Records Collection of the Office of Naval Records and Library, 1691–1945, RG 45, NARA, Washington, DC; Adm. J. G. Walker to Tracy, 15 December 1891, Area Files of the Naval Records Collection, 1775–1910, 1775–1910, RG 45, NARA, Washington, DC.
7. Walker to Tracy, 15 December 1891.
8. Adm. J. G. Walker to Tracy, 20 December 1891, Area Files of the Naval Records Collection, 1775–1910, 1775–1910, RG 45, NARA, Washington, DC; Adm. J. G. Walker to Tracy, telegram, 16 December 1891, Area Files of the Naval Records Collection, 1775–1910, 1775–1910, RG 45, NARA, Washington, DC; Adm. J. G. Walker to Tracy, telegram, 18 December 1891, Area Files of the Naval Records Collection, 1775–1910, 1775–1910, RG 45, NARA, Washington, DC.
9. James R. Soley to Walker, telegram, 21 December 1891, Naval Records Collection of the Office of Naval Records and Library, 1691–1945, RG 45, NARA, Washington, DC.
10. Adm. J. G. Walker to Tracy, 22 December 1891, Naval Records Collection of the Office of Naval Records and Library, 1691–1945, RG 45, NARA, Washington, DC.
11. Benjamin F. Tracy to Walker, telegram, 24 December 1891, Naval Records Collection of the Office of Naval Records and Library, 1691–1945, RG 45, NARA, Washington, DC.
12. Capt. Francis Higginson to Tracy, 6 January 1892, Area Files of the Naval Records Collection, 1775–1910, 1775–1910, RG 45, NARA, Washington, DC; Adm. J. G. Walker to Tracy, 11 January 1892, Area Files of the Naval Records Collection, 1775–1910, 1775–1910, RG 45, NARA, Washington, DC; Adm. J. G. Walker to Tracy, 12 January 1892, Area Files of the Naval Records Collection, 1775–1910, 1775–1910, RG 45, NARA, Washington, DC; Adm. J. G. Walker to Tracy, telegram, 15 January 1892, Area Files of the Naval Records Collection, 1775–1910, 1775–1910, RG 45, NARA, Washington, DC.
13. Benjamin F. Tracy to Gherardi, telegram, 24 December 1891, Naval Records Collection of the Office of Naval Records and Library, 1691–1945, RG 45, NARA, Washington, DC.
14. Commander Belden to Tracy, 2 January 1892, Area Files of the Naval Records Collection, 1775–1910, 1775–1910, RG 45, NARA, Washington, DC.

15. Adm. J. G. Walker to Tracy, 28 January 1892, Area Files of the Naval Records Collection, 1775–1910, 1775–1910, RG 45, NARA, Washington, DC.

16. Adm. J. G. Walker to Tracy, 2 February 1892, Area Files of the Naval Records Collection, 1775–1910, 1775–1910, RG 45, NARA, Washington, DC.

17. Benjamin F. Tracy to Walker, telegram, 13 January 1892, Naval Records Collection of the Office of Naval Records and Library, 1691–1945, RG 45, NARA, Washington, DC.

18. "The Chicago in Port," *New York Times,* 26 May 1892, 4.

19. Rear Adm. Bancroft Gherardi to Walker, 6 February 1892, Records of the Operating Forces, RG 313, NARA, Washington, DC.

20. "You [Rear Admiral Walker] will be pleased to direct . . ." Formal language very clearly and unquestionably denoting a senior/subordinate relationship. Rear Adm. Bancroft Gherardi to Walker, 13 February 1892, Records of the Operating Forces, RG 313, NARA, Washington, DC.

21. Captain Ramsey to Gherardi, telegram, 9 February 1892, Naval Records Collection of the Office of Naval Records and Library, 1691–1945, RG 45, NARA, Washington, DC.

22. Rear Adm. Bancroft Gherardi to SECNAV, telegram, 16 February 1892, Area Files of the Naval Records Collection, 1775–1910, 1775–1910, RG 45, NARA, Washington, DC; "Various Naval Items, 30 Jan 92," *Army and Navy Journal,* 30 January 1892.

23. Adm. J. G. Walker to Tracy, 15 February 1892, Area Files of the Naval Records Collection, 1775–1910, 1775–1910, RG 45, NARA, Washington, DC.

24. Adm. J. G. Walker to Tracy, 30 March 1892, Area Files of the Naval Records Collection, 1775–1910, 1775–1910, RG 45, NARA, Washington, DC.

25. Adm. J. G. Walker to Tracy, telegram, 17 March 1892, Area Files of the Naval Records Collection, 1775–1910, 1775–1910, RG 45, NARA, Washington, DC.

26. Adm. J. G. Walker to Tracy, telegram, 27 April 1892, Area Files of the Naval Records Collection, 1775–1910, 1775–1910, RG 45, NARA, Washington, DC.

27. Lt. R. B. Bradford to Tracy, 5 May 1892, Area Files of the Naval Records Collection, 1775–1910, 1775–1910, RG 45, NARA, Washington, DC.

28. Rear Adm. Bancroft Gherardi to Tracy, 11 March 1892, Area Files of the Naval Records Collection, 1775–1910, 1775–1910, RG 45, NARA, Washington, DC. Cardiff was the main port for export of high-quality coal from Wales. Eureka coal was from Pennsylvania.

29. Rear Adm. Bancroft Gherardi to SECNAV, 8 April 1892, Area Files of the Naval Records Collection, 1775–1910, 1775–1910, RG 45, NARA, Washington, DC.

30. "Going to Memphis," *Atlanta Constitution,* 7 May 1892.

31. Rear Adm. Bancroft Gherardi to SECNAV, telegram, 23 April 1892, Area Files of the Naval Records Collection, 1775–1910, 1775–1910, RG 45, NARA, Washington, DC.

32. Capt. Silas Casey to Tracy, 28 March 1892, Area Files of the Naval Records Collection, 1775–1910, 1775–1910, RG 45, NARA, Washington, DC.

33. Ibid.

34. Capt. Silas Casey to Tracy, 9 April 1892, Area Files of the Naval Records Collection, 1775–1910, 1775–1910, RG 45, NARA, Washington, DC; Capt. Silas Casey to Tracy, 30 April 1892, Area Files of the Naval Records Collection, 1775–1910, 1775–1910, RG 45, NARA, Washington, DC.

35. Commander Winn, Commandant of Naval Station Key West, to Bureau of Navigation and Detail, telegram, 12 May 1892, Area Files of the Naval Records Collection, 1775–1910, 1775–1910, RG 45, NARA, Washington, DC.

36. Rear Adm. Bancroft Gherardi to Tracy, 14 May 1892, Area Files of the Naval Records Collection, 1775–1910, 1775–1910, RG 45, NARA, Washington, DC.

37. Ibid.

38. Rear Adm. Bancroft Gherardi to SECNAV, 21 May 1892, Area Files of the Naval Records Collection, 1775–1910, 1775–1910, RG 45, NARA, Washington, DC; "May Week in Savannah," *Atlanta Constitution,* 11 May 1892.

39. Gherardi to Tracy, 14 May 1892.

40. Rear Adm. Bancroft Gherardi to SECNAV, telegram, 18 May 1892, Area Files of the Naval Records Collection, 1775–1910, 1775–1910, RG 45, NARA, Washington, DC.

41. Gherardi to SECNAV, 21 May 1892.

42. Rear Adm. Bancroft Gherardi to SECNAV, 24 May 1892, Area Files of the Naval Records Collection, 1775–1910, 1775–1910, RG 45, NARA, Washington, DC.

43. Henry Erben to Bureau of Navigation and Detail, telegram, 10 June 1892, Area Files of the Naval Records Collection, 1775–1910, 1775–1910, RG 45, NARA, Washington, DC; Rear Adm. Bancroft Gherardi to SECNAV, 13 June 1892, Area Files of the Naval Records Collection, 1775–1910, 1775–1910, RG 45, NARA, Washington, DC; Captain Weaver, Commandant of Navy Yard Portsmouth, to Bureau of Navigation and Detail, telegram, 11 June 1892, Area Files of the Naval Records Collection, 1775–1910, 1775–1910, RG 45, NARA, Washington, DC.

44. Rear Adm. Bancroft Gherardi to Tracy, 5 July 1892, Area Files of the Naval Records Collection, 1775–1910, 1775–1910, RG 45, NARA, Washington, DC; Rear Adm. Bancroft Gherardi to Tracy, 29 June 1892, Area Files of the Naval Records Collection, 1775–1910, 1775–1910, RG 45, NARA, Washington, DC; "War Ships at New-London," *New York Times,* 5 August 1892.

45. "Army and Navy," *New York Times,* 9 August 1892, 2.

46. Rear Adm. Bancroft Gherardi to SECNAV, telegram, 6 August 1892, Area Files of the Naval Records Collection, 1775–1910, 1775–1910, RG 45, NARA, Washington, DC.

47. Lt. Cdr. W. H. Brownson to SECNAV, telegram, 9 August 1892, Area Files of the Naval Records Collection, 1775–1910, 1775–1910, RG 45, NARA, Washington, DC; "Doings at Bar Harbor," *New York Times,* 18 August 1892, 5.

48. Rear Adm. Bancroft Gherardi to Tracy, 21 August 1892, Area Files of the Naval Records Collection, 1775–1910, 1775–1910, RG 45, NARA, Washington, DC.

49. Rear Adm. Bancroft Gherardi to Tracy, 27 August 1892, Area Files of the Naval Records Collection, 1775–1910, 1775–1910, RG 45, NARA, Washington, DC.

50. Beach and Beach, *From Annapolis to Scapa Flow,* 54.

51. As an example, see Adm. J. G. Walker to SECNAV, telegram, 23 July 1892, Area Files of the Naval Records Collection, 1775–1910, 1775–1910, RG 45, NARA, Washington, DC.

52. "Admiral Walker's Return," *New York Times,* 30 April 1892, 1; "Various Naval Items, 4 Jun 92," *Army and Navy Journal,* 4 June 1892.

53. Adm. J. G. Walker to Tracy, 4 September 1892, Area Files of the Naval Records Collection, 1775–1910, 1775–1910, RG 45, NARA, Washington, DC.

54. Adm. J. G. Walker to Luce, 6 January 1887, Walker Papers.

55. Adm. J. G. Walker to SECNAV, 9 September 1892, Area Files of the Naval Records Collection, 1775–1910, 1775–1910, RG 45, NARA, Washington, DC.

56. "Various Naval Items, 17 Sept 92," *Army and Navy Journal,* 17 September 1892.

57. "Various Naval Items, 1 Oct 92," *Army and Navy Journal,* 1 October 1892; "Various Naval Items, 17 Sept 92."

58. Rear Adm. Bancroft Gherardi to SECNAV, 24 September 1892, Records of Naval Operating Forces, 1849–1980, RG 313, NARA, Washington, DC.

59. J. Otis, *The Boys of '98* (D. Estes: Howard University, 1898), 164.

60. Rear Adm. Bancroft Gherardi to SECNAV, 10 October 1892, Records of Naval Operating Forces, 1849–1980, RG 313, NARA, Washington, DC.

61. Rear Adm. Bancroft Gherardi to SECNAV, 15 October 1892, Records of Naval Operating Forces, 1849–1980, RG 313, NARA, Washington, DC.

62. Rear Adm. Bancroft Gherardi to SECNAV, 5 November 1892, Records of Naval Operating Forces, 1849–1980, RG 313, NARA, Washington, DC.

63. Rear Adm. Bancroft Gherardi to SECNAV, 19 December 1892, Records of Naval Operating Forces, 1849–1980, RG 313, NARA, Washington, DC.

64. Rear Adm. Bancroft Gherardi to SECNAV, 6 December 1892, Records of Naval Operating Forces, 1849–1980, RG 313, NARA, Washington, DC; Rear Adm. Bancroft Gherardi to SECNAV, 10 December 1892, Records of Naval Operating Forces, 1849–1980, RG 313, NARA, Washington, DC.

65. Gherardi to SECNAV, 6 December 1892; Rear Adm. Bancroft Gherardi to SECNAV, 16 November 1892, Records of Naval Operating Forces, 1849–1980, RG 313, NARA, Washington, DC.

66. Gherardi to SECNAV, 10 December 1892.

67. Again, Gherardi simply says that "the four ships were exercised at fleet tactics." The reader is left to decide for himself what that means, although only so many things can be done with four ships. Based on previous squadron work, it is a safe assumption that the four vessels formed in line abreast and column and practiced shifting between those two formations and probably to a line of columns (of two); all this is described in Parker's *Fleet Tactics Under Steam.*

68. Captain Ramsey to Gherardi, telegram, 8 December 1892, Naval Records Collection of the Office of Naval Records and Library, 1691–1945, RG 45, NARA, Washington, DC.

69. Gherardi to SECNAV, 19 December 1892.

70. Rear Adm. Bancroft Gherardi to SECNAV, 24 December 1892, Records of Naval Operating Forces, 1849–1980, RG 313, NARA, Washington, DC.

71. Cdr. Edwin White to Tracy, 8 September 1892, Area Files of the Naval Records Collection, 1775–1910, 1775–1910, RG 45, NARA, Washington, DC; Cdr. Edwin White to Tracy, 15 September 1892, Area Files of the Naval Records Collection, 1775–1910, 1775–1910, RG 45, NARA, Washington, DC; Cdr. Edwin White to Tracy, 16 September 1892, Area Files of the Naval Records Collection, 1775–1910, 1775–1910, RG 45, NARA, Washington, DC.

72. Adm. J. G. Walker to Barker, 10 September 1892, Records of Naval Operating Forces, 1849–1980, RG 313, NARA, Washington, DC.

73. Adm. J. G. Walker to White, 20 September 1892, Records of Naval Operating Forces, 1849–1980, RG 313, NARA, Washington, DC.

74. Adm. J. G. Walker to Crowninshield, 22 September 1892, Records of Naval Operating Forces, 1849–1980, RG 313, NARA, Washington, DC.

75. Adm. J. G. Walker to Tracy, 28 September 1892, Area Files of the Naval Records Collection, 1775–1910, 1775–1910, RG 45, NARA, Washington, DC.

76. Adm. J. G. Walker to Crowninshield, 29 September 1892, Records of Naval Operating Forces, 1849–1980, RG 313, NARA, Washington, DC.

77. Adm. J. G. Walker to Tracy, 19 October 1892, Area Files of the Naval Records Collection, 1775–1910, 1775–1910, RG 45, NARA, Washington, DC; Adm. J. G. Walker to Tracy, 25 October 1892, Area Files of the Naval Records Collection, 1775–1910, 1775–1910, RG 45, NARA, Washington, DC.

78. Rear Adm. Bancroft Gherardi to SECNAV, 23 January 1893, Records of Naval Operating Forces, 1849–1980, RG 313, NARA, Washington, DC.

79. Rear Adm. Bancroft Gherardi to SECNAV, 14 January 1893, Records of Naval Operating Forces, 1849–1980, RG 313, NARA, Washington, DC.

80. Rear Adm. Bancroft Gherardi to Tracy, 18 February 1893, Area Files of the Naval Records Collection, 1775–1910, RG 45, NARA, Washington, DC.

81. Rear Adm. Bancroft Gherardi to SECNAV, 12 March 1893, Records of Naval Operating Forces, 1849–1980, RG 313, NARA, Washington, DC.

82. Shuttered lamps for sending Morse code had not been invented yet.

83. Rear Adm. Bancroft Gherardi to SECNAV, 15 February 1893, Records of Naval Operating Forces, 1849–1980, RG 313, NARA, Washington, DC.

84. "Special Service Squadron," *Army and Navy Journal,* 4 March 1893.

85. Reid R. Badger, *The Great American Fair: The World's Columbian Exposition and American Culture* (Chicago: N. Hall, 1979); T. J. Boisseau, A. M. Markwyn, and R. Rydell, *Gendering the Fair: Histories of Women and Gender at World's Fairs* (Urbana: University of Illinois Press, 2010); R. W. Rydell, *All the World's a Fair: Visions of Empire at American International Expositions, 1876–1916* (Chicago: University of Chicago Press, 1987); R. W. Rydell, *World of Fairs: The Century-of-Progress Expositions* (Chicago: University of Chicago Press, 1993).

86. Gherardi to SECNAV, 12 March 1893; Rear Adm. Bancroft Gherardi to SECNAV, 22 May 1893, Records of Naval Operating Forces, 1849–1980, RG 313, NARA, Washington, DC.

87. "Admiral Gherardi's Fleet," *New York Times,* 26 February 1893, 3; "Cool from Admiral Walker," *New York Times,* 10 January 1893, 1; "Walker Not Satisfied," *New York Times,* 25 February 1893, 10.

88. Rear Adm. Bancroft Gherardi to Herbert, 24 March 1893, Area Files of the Naval Records Collection, 1775–1910, RG 45, NARA, Washington, DC; Rear Adm. Bancroft Gherardi to SECNAV, 24 March 1893, Records of Naval Operating Forces, 1849–1980, RG 313, NARA, Washington, DC.

89. Rear Adm. Bancroft Gherardi to Herbert, 29 March 1893, Area Files of the Naval Records Collection, 1775–1910, RG 45, NARA, Washington, DC.

90. William Folger to Tracy, 27 December 1892, Area Files of the Naval Records Collection, 1775–1910, RG 45, NARA, Washington, DC.

91. Department of the Navy, *Annual Report of the Secretary of the Navy,* 1893, vol. 1, pp. 181–196.

92. Henry Erben to SECNAV, telegram, 30 March 1893, Area Files of the Naval Records Collection, 1775–1910, RG 45, NARA, Washington, DC; Rear Adm. Bancroft Gherardi

to Herbert, telegram, 31 March 1893, Area Files of the Naval Records Collection, 1775–1910, RG 45, NARA, Washington, DC.

93. Rear Adm. Bancroft Gherardi to Benham, 31 March 1893, memorandum, Records of Naval Operating Forces, 1849–1980, RG 313, NARA, Washington, DC.

94. "New Code of Naval Signals," *New York Times,* 26 January 1893, 6.

95. Gherardi to SECNAV, 22 May 1893, 3; Rear Adm. Bancroft Gherardi to Walker, 3 April 1893, Records of Naval Operating Forces, 1849–1980, RG 313, NARA, Washington, DC.

96. Rear Adm. Bancroft Gherardi to Herbert, 11 April 1893, Area Files of the Naval Records Collection, 1775–1910, RG 45, NARA, Washington, DC; Rear Adm. Bancroft Gherardi to Herbert, telegram, 11 April 1893, Area Files of the Naval Records Collection, 1775–1910, RG 45, NARA, Washington, DC.

97. Rear Adm. Bancroft Gherardi to the Naval Review Fleet, memorandum, 7 April 1893, Records of Naval Operating Forces, 1849–1980, RG 313, NARA, Washington, DC.

98. Hilary A. Herbert, "The Lesson of the Naval Review," *North American Review* 156, no. 439 (1893).

99. "The International Naval Review," *Army and Navy Journal,* 29 April 1893.

100. Modern events like this are done with a heavy reliance on e-mail, exchanges of officers by helicopter, bridge-to-bridge radio, and other modern means of communication—and they are still difficult. That this formation could be brought together and maneuver at a uniform speed while keeping station, all by way of signal flags, has to be recognized as a truly impressive monument not only to Rear Admiral Gherardi's organizational skills but also to the seamanship of the various assembled navies and the advances made by the U.S. Navy.

101. Rear Adm. Bancroft Gherardi to Herbert, 15 April 1893, Area Files of the Naval Records Collection, 1775–1910, RG 45, NARA, Washington, DC; Rear Adm. Bancroft Gherardi, "Programme for Naval Review 1893," 15, Records of Naval Operating Forces, 1849–1980, RG 313, NARA, Washington, DC.

102. So even though Gherardi was against it, Mr. Miller of the New York State Militia ended up getting his way in the end. See Rear Adm. Bancroft Gherardi to Miller, 12 April 1893, Records of Naval Operating Forces, 1849–1980, RG 313, NARA, Washington, DC.

103. "Admiral Gherardi's Command," *New York Times,* 19 April 1893, 4; "After the Naval Review," *New York Times,* 3 May 1893, 2; "Seniority Should Prevail," *New York Times,* 26 April 1893, 4.

104. "Gherardi to Leave the Sea," *New York Times,* 20 May 1893, 1.

105. "After the Naval Review."

106. "How Our Navy Is Distributed," *New York Times,* 28 December 1893, 3.

107. "After the Naval Review."

108. "How Our Navy Is Distributed."

109. "Blame for Admiral Tryon," *New York Times,* 7 July 1893, 10.

110. Rear Adm. A. E. K. Benham to Herbert, 31 October 1893, Area Files of the Naval Records Collection, 1775–1910, RG 45, NARA, Washington, DC.

111. Ibid.

112. Department of the Navy, *Annual Report of the Secretary of the Navy,* 1894, vol. 1, p. 23.

Chapter 5. Luce's Vision Realized

1. Rear Adm. R. W. Meade to Herbert, 4 January 1895, Area Files of the Naval Records Collection, 1775–1910, RG 45, NARA, Washington, DC; "The North Atlantic Squadron," *New York Times,* 21 January 1895, 12.

2. Rear Adm. R. W. Meade to Herbert, 1 January 1895, Area Files of the Naval Records Collection, 1775–1910, RG 45, NARA, Washington, DC.

3. "Starts for the Rendezvous," *New York Times,* 11 January 1895, 1.

4. Timothy S. Wolters makes this point in Wolters, "Recapitalizing the Fleet: A Material Analysis of Late-Nineteenth-Century U.S. Naval Power," *Technology and Culture* 52, no. 1 (2011): 24.

5. Rear Adm. R. W. Meade to Cromwell, 17 January 1895, Area Files of the Naval Records Collection, 1775–1910, RG 45, NARA, Washington, DC; Rear Adm. R. W. Meade to Herbert, 17 January 1895, Area Files of the Naval Records Collection, 1775–1910, RG 45, NARA, Washington, DC.

6. "I regret to say that I did not expect very much from Captain Cromwell when I heard he had been appointed to the ATLANTA," Meade to Herbert, 17 January 1895.

7. Rear Adm. R. W. Meade to Herbert, 30 January 1895, Area Files of the Naval Records Collection, 1775–1910, RG 45, NARA, Washington, DC.

8. Rear Adm. R. W. Meade to Herbert, 1 March 1895, Area Files of the Naval Records Collection, 1775–1910, RG 45, NARA, Washington, DC.

9. The Battle of the Yalu in 1894 yielded several important lessons learned, among them the danger of flying splinters from interior woodwork shattered by enemy shells impacting on the armored exterior surface of a ship. Admiral Meade obviously had this in mind when he criticized the internal arrangements of *New York.*

10. Rear Adm. R. W. Meade to Herbert, 6 February 1895, Area Files of the Naval Records Collection, 1775–1910, RG 45, NARA, Washington, DC.

11. Rear Adm. R. W. Meade to Herbert, 23 February 1895, Enclosure 2, Area Files of the Naval Records Collection, 1775–1910, RG 45, NARA, Washington, DC.

12. Meade to Herbert, 1 March 1895.

13. Rear Adm. R. W. Meade to Herbert, 3 March 1895, Area Files of the Naval Records Collection, 1775–1910, RG 45, NARA, Washington, DC.

14. Ibid.

15. Evans, *Sailor's Log,* 364.

16. A formal request from the governor arrived about ten minutes after the boats cast off. Robley Evans notes in his autobiography that the delay caused "much property to be lost." In Meade's defense, the North Atlantic Squadron had not had great luck with putting landing parties ashore over the previous couple of years—*Chicago's* 1894 debacle with drunken Marines and *Atlanta's* strained relationship with the new revolutionary government of Nicaragua after putting troops ashore being two examples. See chapter 4. Meade probably had these things in mind when hesitating to land assistance.

17. Evans, *Sailor's Log,* 365.

18. "Colombia's Stubborn Rebels," *New York Times,* 7 March 1895, 5.

19. Rear Adm. R. W. Meade to Herbert, 7 March 1895, Area Files of the Naval Records Collection, 1775–1910, RG 45, NARA, Washington, DC.

20. Capt. Merrill Miller to Herbert, 19 March 1895, Area Files of the Naval Records Collection, 1775–1910, RG 45, NARA, Washington, DC.

21. Heureaux's government was heavily in debt to several European countries, France included.

22. "To See the St. Paul Launched," *New York Times*, 25 March 1895, 1.

23. Rear Adm. R. W. Meade to Herbert, 10 April 1895, Area Files of the Naval Records Collection, 1775–1910, RG 45, NARA, Washington, DC; Rear Adm. R. W. Meade to Herbert, 11 April 1895, Area Files of the Naval Records Collection, 1775–1910, RG 45, NARA, Washington, DC; Rear Adm. R. W. Meade to Wadleigh, 10 April 1895, Area Files of the Naval Records Collection, 1775–1910, RG 45, NARA, Washington, DC.

24. Secretary of the Navy Hilary A. Herbert to Meade, 19 April 1895, Area Files of the Naval Records Collection, 1775–1910, RG 45, NARA, Washington, DC; Meade to Herbert, 10 April 1895.

25. Rear Adm. R. W. Meade to Herbert, 24 April 1895, Area Files of the Naval Records Collection, 1775–1910, RG 45, NARA, Washington, DC.

26. Evans, *Sailor's Log,* 364.

27. Meade to Herbert, 7 March 1895; Rear Adm. R. W. Meade to Herbert, 19 April 1895, Area Files of the Naval Records Collection, 1775–1910, RG 45, NARA, Washington, DC. Meade recommended simply seizing the isthmus and sending two regiments of "colored troops" to hold each end of it. There is no explanation of why he felt that African Americans should be used, but his letters consistently remark about the unhealthiness of the oppressive heat, and he probably felt—as many did in 1895—that African Americans were more suited for work in tropical climates than whites.

28. Meade to Herbert, 19 April 1895.

29. Rear Adm. R. W. Meade to SECNAV, telegram, 20 April 1895, Area Files of the Naval Records Collection, 1775–1910, RG 45, NARA, Washington, DC.

30. Meade to Herbert, 24 April 1895.

31. Rear Adm. R. W. Meade to Herbert, 28 April 1895, Area Files of the Naval Records Collection, 1775–1910, RG 45, NARA, Washington, DC.

32. Rear Adm. R. W. Meade to Herbert, 6 May 1895, Area Files of the Naval Records Collection, 1775–1910, RG 45, NARA, Washington, DC.

33. Capt. Henry Glass to Herbert, 13 May 1895, Area Files of the Naval Records Collection, 1775–1910, RG 45, NARA, Washington, DC.

34. "Herbert Shows up Meade," *New York Times*, 15 May 1895, 3; "Meade's Ensign Down," *New York Tribune*, 10 May 1895.

35. "Admiral Meade May Retire," *New York Times*, 5 May 1895, 11.

36. "Meade's Ensign Down."

37. "Talk About His Successor," *New York Times*, 14 May 1895, 3.

38. Rear Adm. F. M. Bunce to Herbert, 11 July 1895, Area Files of the Naval Records Collection, 1775–1910, RG 45, NARA, Washington, DC.

39. Rear Adm. F. M. Bunce to Wise, 16 July 1895, Area Files of the Naval Records Collection, 1775–1910, RG 45, NARA, Washington, DC.

40. "Atlanta to Be Thoroughly Overhauled," *New York Times*, 15 August 1895, 9.

41. "Atlantic Squadron at Newport," *New York Times*, 9 August 1895, 1.

42. "Fleet Off for Newport," *New York Times*, 8 August 1895, 9.

43. "The North Atlantic Squadron," *New York Times,* 14 August 1895, 1

44. Rear Adm. F. M. Bunce to Herbert, 14 August 1895, Area Files of the Naval Records Collection, 1775–1910, RG 45, NARA, Washington, DC; "Land Drill of Sailors in Newport," *New York Times,* 10 August 1895, 1; "North Atlantic Squadron."

45. "Fleet Off for Newport."

46. "Squadron Sails for Bar Harbor," *New York Times*, 16 August 1895, 1.

47. Rear Adm. F. M. Bunce to Herbert, 29 August 1895, Area Files of the Naval Records Collection, 1775–1910, RG 45, NARA, Washington, DC.

48. "Waiting for the Atlantic Squadron," *New York Times*, 31 July 1895, 6.

49. "Bar Harbor," *New York Times,* 16 August 1895, 3; "Grand Naval Reception," *New York Times,* 21 August 1895, 3.

50. "No Manoeuvres Necessary," *New York Times*, 22 August 1895, 4.

51. Bunce to Herbert, 29 August 1895.

52. Ibid; "Movements of the White Squadron," *New York Times,* 29 August 1895, 1.

53. "Topics of the Times," *New York Times*, 22 April 1897, 6.

54. "North Atlantic Squadron at Boston," *New York Times*, 31 August 1895, 12.

55. Ship descriptions from Alden, *American Steel Navy,* 31–32.

56. Rear Adm. F. M. Bunce to Crowninshield, 20 December 1895, Area Files of the Naval Records Collection, 1775–1910, RG 45, NARA, Washington, DC.

57. Rear Adm. F. M. Bunce to Herbert, 23 September 1895, Area Files of the Naval Records Collection, 1775–1910, RG 45, NARA, Washington, DC.

58. See chapter 1, figure 1, for definitions of these formations.

59. Rear Adm. F. M. Bunce to Ramsay, 4 October 1895, Area Files of the Naval Records Collection, 1775–1910, RG 45, NARA, Washington, DC.

60. Main battery: the largest-caliber, primary weapons of a warship, to be used against an armored peer opponent. Secondary battery: smaller-caliber guns, which typically had a higher rate of fire, designed to attack unarmored sections of a peer opponent, such as the bridge, as well as protect the ship against smaller threats, such as torpedo boats. See Tucker, *Handbook,* 161.

61. Rear Adm. F. M. Bunce, circular order, Bunce to North Atlantic Squadron, 29 September 1895, Area Files of the Naval Records Collection, 1775–1910, RG 45, NARA, Washington, DC.

62. Rear Adm. F. M. Bunce to Ramsay, 1 November 1895, Area Files of the Naval Records Collection, 1775–1910, RG 45, NARA, Washington, DC.

63. To be able to better judge the fall of a shot, without confusing which battery it was coming from.

64. Rear Adm. F. M. Bunce to Ramsay, 26 October 1895, Area Files of the Naval Records Collection, 1775–1910, RG 45, NARA, Washington, DC.

65. Rear Adm. F. M. Bunce to Ramsay, 1 December 1895, Area Files of the Naval Records Collection, 1775–1910, RG 45, NARA, Washington, DC.

66. Rear Adm. F. M. Bunce to Wise, 16 December 1895, Area Files of the Naval Records Collection, 1775–1910, RG 45, NARA, Washington, DC.

67. Secretary of the Navy Hilary A. Herbert to Bunce, 31 October 1895, Area Files of the Naval Records Collection, 1775–1910, RG 45, NARA, Washington, DC.

68. Rear Adm. F. M. Bunce to Ramsay, 9 December 1895, Area Files of the Naval Records Collection, 1775–1910, RG 45, NARA, Washington, DC.

69. "Utah Wants Her Statehood," *New York Times*, 17 December 1895, 3.

70. Rear Adm. F. M. Bunce to Herbert, 29 December 1895, Area Files of the Naval Records Collection, 1775–1910, RG 45, NARA, Washington, DC; Rear Adm. F. M. Bunce to Ramsay, 20 December 1895, Area Files of the Naval Records Collection, 1775–1910, RG 45, NARA, Washington, DC.

71. "New Year's on the Squadron," *New York Times*, 3 January 1896, 9.

72. Herrick, *American Naval Revolution*, 202.

73. Quoted in Herring, *From Colony to Superpower*, 307.

74. "English Journals Facetious," *New York Times*, 6 January 1896, 5; "Not Unprepared for War," *New York Times*, 19 December 1895, 1.

75. "Bunce's Fleet to Be Held," *New York Times*, 23 December 1895, 2.

76. "Claims against Turkey," *New York Times*, 7 January 1896, 5; "Threatening Fleet's Force," *New York Times*, 7 January 1896, 5.

77. "Will Not Go to Turkey," *New York Times*, 13 January 1896, 2.

78. Rear Adm. F. M. Bunce to Sands, 4 February 1896, Area Files of the Naval Records Collection, 1775–1910, RG 45, NARA, Washington, DC.

79. Rear Adm. F. M. Bunce to Ramsay, 14 April 1896, Area Files of the Naval Records Collection, 1775–1910, RG 45, NARA, Washington, DC.

80. Rear Adm. F. M. Bunce to Ramsay, 29 April 1896, Area Files of the Naval Records Collection, 1775–1910, RG 45, NARA, Washington, DC.

81. Rear Adm. F. M. Bunce to Stirling, 30 April 1896, Area Files of the Naval Records Collection, 1775–1910, RG 45, NARA, Washington, DC.

82. Rear Adm. F. M. Bunce to Herbert, 17 June 1896, Area Files of the Naval Records Collection, 1775–1910, RG 45, NARA, Washington, DC.

83. Rear Adm. F. M. Bunce to Ramsay, 28 July 1896, Area Files of the Naval Records Collection, 1775–1910, RG 45, NARA, Washington, DC.

84. Secretary of the Navy Hilary A. Herbert to Mansfield, 25 May 1896, Area Files of the Naval Records Collection, 1775–1910, RG 45, NARA, Washington, DC.

85. "Admiral Bunce at Sea," *New York Times*, 2 August 1896, 9; "The North Atlantic Squadron," *New York Times*, 18 June 1896, 9.

86. Rear Adm. F. M. Bunce to Ramsay, 9 August 1896, Area Files of the Naval Records Collection, 1775–1910, RG 45, NARA, Washington, DC.

87. Subcaliber target practice: a method of installing what was essentially a rifle along the barrel of a cannon. The gun crew would go through all the correct motions to simulate loading the cannon, then fire the rifle to check proper sighting and alignment of the gun. This saved expenditure of powder and shells.

88. Bunce to Ramsay, 9 August 1896.

89. On later torpedo tactics, see Epstein, "No One Can Afford to Say," 491–520.

90. N. Friedman, *U.S. Battleships: An Illustrated Design History* (Annapolis: Naval Institute Press, 1985), 25–29.

91. Rear Adm. F. M. Bunce to Rodgers, 14 August 1896, Area Files of the Naval Records Collection, 1775–1910, RG 45, NARA, Washington, DC.

92. Rear Adm. F. M. Bunce to Ramsay, 23 August 1896, Area Files of the Naval Records Collection, 1775–1910, RG 45, NARA, Washington, DC.

93. Rear Adm. F. M. Bunce to Ramsay, 15 August 1896, Area Files of the Naval Records Collection, 1775–1910, RG 45, NARA, Washington, DC; Rear Adm. F. M. Bunce, to Ramsay, telegram, 23 August 1896, Area Files of the Naval Records Collection, 1775–1910, RG 45, NARA, Washington, DC; "White Ships Here Again," *New York Times,* 24 August 1896, 8.

94. Rear Adm. F. M. Bunce to Ramsay, 24 August 1896, Area Files of the Naval Records Collection, 1775–1910, RG 45, NARA, Washington, DC.

95. Rear Adm. F. M. Bunce to Stirling, 27 August 1896, Area Files of the Naval Records Collection, 1775–1910, RG 45, NARA, Washington, DC.

96. Rear Adm. F. M. Bunce to Herbert, 29 August 1896, Area Files of the Naval Records Collection, 1775–1910, RG 45, NARA, Washington, DC.

97. Rear Adm. F. M. Bunce to Ramsay, 5 September 1896, Area Files of the Naval Records Collection, 1775–1910, RG 45, NARA, Washington, DC.

98. Rear Adm. F. M. Bunce to Ramsay, 14 September 1896, Area Files of the Naval Records Collection, 1775–1910, RG 45, NARA, Washington, DC.

99. Rear Adm. F. M. Bunce to Ramsay, 19 September 1896, Area Files of the Naval Records Collection, 1775–1910, RG 45, NARA, Washington, DC.

100. Rear Adm. F. M. Bunce to Ramsay, telegram, 1 October 1896, Area Files of the Naval Records Collection, 1775–1910, RG 45, NARA, Washington, DC.

101. Rear Adm. F. M. Bunce to Herbert, 7 October 1896, Area Files of the Naval Records Collection, 1775–1910, RG 45, NARA, Washington, DC.

102. Evans, *Sailor's Log*, 399.

103. Rear Adm. F. M. Bunce to Ramsay, 14 October 1896, Area Files of the Naval Records Collection, 1775–1910, RG 45, NARA, Washington, DC.

104. Note the difference in the titles. By 1896 it is assumed that all squadron tactics will take place "under steam," rendering the phrase superfluous.

105. Bunce to Ramsay, 14 October 1896 Rear Adm. F. M. Bunce, *Squadron Tactics,* 1896. Located in *Naval Historical Collection*, U.S. Naval War College, Newport, RI.

106. Rear Adm. F. M. Bunce to Ramsay, 1 December 1896, Area Files of the Naval Records Collection, 1775–1910, RG 45, NARA, Washington, DC.

107. Rear Adm. F. M. Bunce to Herbert, 8 December 1896, Area Files of the Naval Records Collection, 1775–1910, RG 45, NARA, Washington, DC. "I have to acknowledge the receipt . . . authorizing me to send the vessels of the squadron under my command to Hampton Roads whenever it is deemed proper."

108. Rear Adm. F. M. Bunce to Ramsay, 10 December 1896, Area Files of the Naval Records Collection, 1775–1910, RG 45, NARA, Washington, DC.

109. "Will Blockade Charleston," *New York Times*, 9 January 1897, 2.

110. At the risk of engaging in teleology, the fact that the squadron's major contribution to the Spanish-American-Cuban War the following year was a blockade action in foreign waters speaks for itself.

111. Robley Evans, in his memoirs, devotes an entire chapter to complaining about how difficult it was to get *Indiana* into Port Royal to utilize the dry dock the Navy had

built there. That episode signaled the beginning of the end of Port Royal as a major naval station.

112. Rear Adm. F. M. Bunce to North Atlantic Squadron, 2 February 1897, Area Files of the Naval Records Collection, 1775–1910, 1775–1910, RG 45, NARA, Washington, DC.

113. "The Marblehead's Injured," *New York Times*, 10 February 1897, 3.

114. "Welcomed to Charleston," *New York Times*, 11 February 1897, 3.

115. Bunce's report is unclear as to the nature of these torpedo exercises, but the assumption throughout the report is that the ships are at anchor. Bunce makes later reference to stationary firing exercises with "such guns as could be brought to bear." It is likely that the torpedo crews were exercised on whichever broadside tubes were facing out to sea while the ship remained anchored.

116. "Warships at Charleston," *New York Times,* 13 February 1897, 2.

117. Rear Adm. F. M. Bunce to Herbert, 23 February 1897, 3, Area Files of the Naval Records Collection, 1775–1910, 1775–1910, RG 45, NARA, Washington, DC.

118. "The Blockade Is Broken," *New York Times.* 14 February 1897, 3.

119. Bunce to Herbert, 23 February 1897.

120. Rear Adm. F. M. Bunce to Rodgers, 2 March 1897, Area Files of the Naval Records Collection, 1775–1910, RG 45, NARA, Washington, DC.

121. Rear Adm. F. M. Bunce to Glass, 7 March 1897, Area Files of the Naval Records Collection, 1775–1910, RG 45, NARA, Washington, DC; Rear Adm. F. M. Bunce to Ramsay, 1 March 1897, Area Files of the Naval Records Collection, 1775–1910, RG 45, NARA, Washington, DC; Rear Adm. F. M. Bunce to Ramsay, 23 February 1897, Area Files of the Naval Records Collection, 1775–1910, RG 45, NARA, Washington, DC.

122. Cdr. E. T. Strong to Ramsay, 23 December 1896, Area Files of the Naval Records Collection, 1775–1910, RG 45, NARA, Washington, DC.

123. Rear Adm. F. M. Bunce, circular order, Bunce to North Atlantic Squadron, 5 March 1897, Area Files of the Naval Records Collection, 1775–1910, RG 45, NARA, Washington, DC.

124. Bunce to Glass, 7 March 1897. A "section" is a pair of warships.

125. Rear Adm. F. M. Bunce to Ramsay, 20 March 1897, Area Files of the Naval Records Collection, 1775–1910, RG 45, NARA, Washington, DC; Capt. W. C. Wise to SECNAV, 12 March 1897, Area Files of the Naval Records Collection, 1775–1910, RG 45, NARA, Washington, DC.

126. Evans, *Sailor's Log,* 398.

127. "Bunce Now in Command," *New York Times,* 2 May 1897, 23.

128. Cogar, *Dictionary of Admirals of the U.S. Navy,* 167–168.

129. Louis A. Perez Jr., *On Becoming Cuban: Identity, Nationality, and Culture* (Chapel Hill: University of North Carolina Press, 1999), 99–100.

130. Herring, *From Colony to Superpower,* 309–310.

131. Rear Adm. Robley Evans credits him with drilling hard to prepare for war. See Evans, *Sailor's Log.*

132. Department of the Navy, *Annual Report of the Secretary of the Navy,* 1897, vol. 1, p. 230.

133. Rear Adm. Montgomery Sicard to Long, 15 June 1897, Area Files of the Naval Records Collection, 1775–1910, RG 45, NARA, Washington, DC.

134. Theodore Roosevelt to Sicard, 17 June 1897, Area Files of the Naval Records Collection, 1775–1910, RG 45, NARA, Washington, DC.

135. T. Roosevelt, *The Works of Theodore Roosevelt: American Ideals, with a Biographical Sketch by F. V. Greene: Administration* (New York: P. F. Collier, 1897), 277.

136. "Squadron Manoeuvres," *New York Times*, 30 June 1897, 4.

137. "A Dinner to Admiral Sicard," *New York Times*, 27 August 1897, 7; Rear Adm. Montgomery Sicard to Ramsay, telegram, 30 August 1897, Area Files of the Naval Records Collection, 1775–1910, RG 45, NARA, Washington, DC; "Society at Bar Harbor," *New York Times,* 29 August 1897, 13.

138. Rear Adm. Montgomery Sicard to Ramsay, telegram, 12 September 1897, Area Files of the Naval Records Collection, 1775–1910, RG 45, NARA, Washington, DC.

139. Rear Adm. Montgomery Sicard to Ramsay, 20 September 1897, Area Files of the Naval Records Collection, 1775–1910, RG 45, NARA, Washington, DC.

140. Theodore Roosevelt to Sicard, 21 September 1897, Area Files of the Naval Records Collection, 1775–1910, RG 45, NARA, Washington, DC.

141. Rear Adm. Montgomery Sicard to Long, 23 September 1897, Area Files of the Naval Records Collection, 1775–1910, RG 45, NARA, Washington, DC.

142. Rear Adm. Montgomery Sicard to Long, 2 October 1897, Area Files of the Naval Records Collection, 1775–1910, RG 45, NARA, Washington, DC.

143. Rear Adm. Montgomery Sicard to Ramsay, 14 October 1897, Area Files of the Naval Records Collection, 1775–1910, RG 45, NARA, Washington, DC.

144. Ibid.

145. Rear Adm. Montgomery Sicard to Ramsay, 24 October 1897, Area Files of the Naval Records Collection, 1775–1910, RG 45, NARA, Washington, DC.

146. Rear Adm. Montgomery Sicard to Wise, 10 November 1897, Area Files of the Naval Records Collection, 1775–1910, RG 45, NARA, Washington, DC.

147. Herrick, *American Naval Revolution,* 165–166.

148. Rear Adm. Montgomery Sicard to Bradford, telegram, 7 November 1987, Area Files of the Naval Records Collection, 1775–1910, RG 45, NARA, Washington, DC.

149. *Chicago,* while John G. Walker's flagship, was said to have an excellent baseball team.

150. John D. Long to Sicard, 3 December 1897, Area Files of the Naval Records Collection, 1775–1910, RG 45, NARA, Washington, DC.

151. Although they would deploy without him. Sicard was condemned by medical survey in late 1897 and had to be replaced by W. T. Sampson right before the outbreak of hostilities.

152. Rear Adm. Montgomery Sicard to CO of Brooklyn, 28 December 1897, Area Files of the Naval Records Collection, 1775–1910, RG 45, NARA, Washington, DC.

153. Rear Adm. Montgomery Sicard to Sigsbee, 8 December 1897, Area Files of the Naval Records Collection, 1775–1910, RG 45, NARA, Washington, DC.

154. Evans, *Sailor's Log,* 404.

Epilogue

1. The following account draws from Chaplain William G. Cassard, USN, ed., *Battleship* Indiana *and Her Part in the Spanish-American War* (New York: Everett B. Mero, 1898); A. B. Feuer, *The Spanish-American War at Sea: Naval Action in the Atlantic* (Westport, CT: Praeger, 1995); W. A.

M. Goode, ed., *With Sampson through the War: Being an Account of the Naval Operations of the North Atlantic Squadron During the Spanish-American War of 1898* (New York: Doubleday and McClure, 1899); Jim Leeke, *Manila and Santiago: The New Steel Navy in the Spanish-American War* (Annapolis: U.S. Naval Institute Press, 2009); John D. Long, *The New American Navy,* 2 vols. (New York: Outlook, 1903); Craig L. Symonds, *The Naval Institute Historical Atlas of the U.S. Navy* (Annapolis: Naval Institute Press, 1995).

2. Morris Janowitz, *The Professional Soldier: A Social and Political Portrait* (New York: Free Press, 1971).

3. Long, *New American Navy,* 1:209–213.

4. Ibid. 1:276.

5. That is, in the opposite direction from the heading of the escaping Spanish ships.

6. The ratio of hits to shells fired at Santiago was something on the order of 1.3 percent. Leeke, *Manila and Santiago,* 153; Symonds, *Naval Institute Historical Atlas,* 114.

7. Quoted in Leeke, *Manila and Santiago,* 141.

8. Long, *New American Navy,* 47.

9. Jerry W. Jones, *U.S. Battleship Operations in World War I* (Annapolis: Naval Institute Press, 1998).

10. For instance, C. C. Felker, *Testing American Sea Power: U.S. Navy Strategic Exercises, 1923–1940* (College Station: Texas A&M University Press, 2006).

11. Parker, "Our Fleet Maneuvers in the Bay of Florida."

12. Buhl, "Maintaining an American Navy."

13. Love, *History of the U.S. Navy.*

14. Roberts, "Indicator of Informal Empire."

15. Plesur, *America's Outward Thrust;* Milton Plesur, *Creating an American Empire, 1865–1914,* ed. A. S. Eisenstadt, Major Issues in American History (New York: Pitman, 1971); William Appleman Williams, *The Roots of the Modern American Empire: A Study of the Growth and Shaping of Social Consciousness in a Marketplace Society* (New York: Random House, 1969).

16. W. A. Owens, *High Seas: The Naval Passage to an Uncharted World* (Annapolis: Naval Institute Press, 1995); Adm. J. Paul Reason, *Sailing New Seas,* ed. Patricia A. Goodrich, Newport Papers series (Newport, RI: Naval War College Press, 1998).

BIBLIOGRAPHY

Primary Sources

National Archives and Records Administration (NARA)

RG 19, Records of the Bureau of Construction and Repair

RG 24, Records of the Bureau of Naval Personnel

RG 45, Records Collection of the Office of Naval Records and Library, 1775–1910

RG 74, Records of the Bureau of Ordnance

RG 313, Records of the Operating Forces

Naval Historical Collection, Naval War College, Newport, RI

RG 4, Naval War College Publications

RG 8, Intelligence and Technical Archives

RG 14, Faculty and Staff Presentations

RG 15, Lectures

Naval Historical Foundation Collection, Library of Congress

Papers of Hamilton Fish

Papers of Stephen Bleeker Luce

Papers of David Dixon Porter

Papers of John Grimes Walker

USS *Olympia,* Independence Seaport Museum, Philadelphia

Handbook of the USS "OLYMPIA," Flagship, Asiatic Station

Very, Lt. E. W., USN, *Report on Torpedo-Boats for Coast Defense* (Washington, DC: GPO, 1884.)

Secondary Sources

Newspapers and Journals

Army and Navy Journal

Atlanta Constitution

Baltimore Sun

New York Herald

New York Times

New York Tribune

U.S. Naval Institute *Proceedings*

The Washington Post

Books

Abbott, Andrew. *The System of Professions: An Essay on the Division of Expert Labor.* Chicago: University of Chicago Press, 1988.

Abrahamson, James L. *America Arms for a New Century: The Making of a Great Military Power.* New York: Free Press, 1981.

Abrahamsson, Bengt. *Military Professionalization and Political Power.* Edited by Morris Janowitz, Charles C. Moskos, and Sam C. Sarkesian. Sage Series on Armed Forces and Society. Beverly Hills, CA: Sage Publications, 1972.

Albion, Robert G. *Makers of Naval Policy, 1798–1947.* Annapolis: U.S. Naval Institute, 1980.

Alden, John D. *The American Steel Navy: A Photographic History of the U.S. Navy from the Introduction of the Steel Hull in 1883 to the Cruise of the Great White Fleet, 1907–1909.* Annapolis: U.S. Naval Institute, 1989.

Ammen, Daniel. *The Old Navy and the New.* Philadelphia, 1891.

Anderson, Fred, and Andrew Cayton. *The Dominion of War: Empire and Liberty in North America, 1400–2000.* New York: Penguin, 2005.

Badger, Reid R. *The Great American Fair: The World's Columbian Exposition and American Culture.* Chicago: N. Hall, 1979.

Baer, George W. *One Hundred Years of Sea Power: The U.S. Navy, 1890–1990.* Stanford, CA: Stanford University Press, 1994.

Ballard, G. A. *The Black Battlefleet.* Annapolis: Naval Institute Press, 1980.

Barker, Rear Adm. Albert S. *Everyday Life in the Navy.* Boston: Gorham Press, 1928.

Baudry, A. *The Naval Battle: Studies of the Tactical Factors.* London: Hugh Rees, 1914.

Beach, Edward L., Sr., and Edward L. Beach Jr. *From Annapolis to Scapa Flow: The Autobiography of Edward L. Beach Sr.* Annapolis: Naval Institute Press, 2003.

Bederman, Gail. *Manliness and Civilization: A Cultural History of Gender and Race in the United States, 1880–1917.* Women in Culture and Society series. Chicago: University of Chicago Press, 1995.

Beeler, John F. *British Naval Policy in the Gladstone-Disraeli Era, 1866–1880.* Stanford, CA: Stanford University Press, 1997.

Beisner, Robert L. *From the Old Diplomacy to the New, 1865–1900.* Edited by John Hope Franklin and Abraham S. Eisenstadt. The Crowell American History Series. New York: Thomas Y. Crowell, 1975.

Bennett, Passed Assistant Engineer Frank M., USN. *The Steam Navy: A History of the Growth of the Steam Vessel of War in the U.S. Navy and of the Naval Engineer Corps.* Pittsburgh: Warren, 1896.

Bigelow, Donald Nevius. *William Conant Church and the Army and Navy Journal.* Edited by The Faculty of Political Science of Columbia University. Studies in History, Economics, and Public Law. New York: Columbia University Press, 1952.

Black, Jeremy. *Rethinking Military History.* New York: Routledge, 2004.

———. *War and the World: Military Power and the Fate of Continents, 1450–2000.* New Haven, CT: Yale University Press, 1998.

Bledstein, Burton. *The Culture of Professionalism: The Middle Class and the Development of Higher Education in America.* New York: W. W. Norton, 1976.

Blight, D. W. *Race and Reunion: The Civil War in American Memory.* Cambridge: Belknap Press of Harvard University Press, 2001.

Boisseau, T. J., A. M. Markwyn, and R. Rydell. *Gendering the Fair: Histories of Women and Gender at World's Fairs.* Urbana: University of Illinois Press, 2010.

Boot, Max. *The Savage Wars of Peace: Small Wars and the Rise of American Power.* New York: Basic Books, 2002.

Bradford, Richard H. *The Virginius Affair.* Boulder: Colorado Associated University Press, 1980.

Brands, H. W. *American Colossus: The Triumph of Capitalism, 1865–1900.* New York: Doubleday, 2010.

Brassey, T. A. *The British Navy: Its Strengths, Resources, and Administration.* London, 1882–83.

Brendon, Piers. *The Decline and Fall of the British Empire, 1781–1997.* New York: Alfred A. Knopf, 2008.

Brodie, Bernard. *Seapower in the Machine Age.* Princeton, NJ: Princeton University Press, 1943.

Browning, Robert M. *From Cape Charles to Cape Fear: The North Atlantic Blockading Squadron During the Civil War.* Tuscaloosa: University of Alabama Press, 1993.

———. *Success Is All that Was Expected: The South Atlantic Blockading Squadron During the Civil War.* Dulles, VA: Brassey's, 2002.

Brownson, Willard H. *From Frigate to Dreadnaught.* Sharon, CT: King House, 1973.

Brzezinski, Zbigniew. *Strategic Vision: America and the Crisis of Global Power.* New York: Basic Books, 2012.

Budiansky, Stephen. *Perilous Fight: America's Intrepid War with Britain on the High Seas, 1812–1815.* New York: Alfred A. Knopf, 2010.

Campbell, Charles S. *The Transformation of American Foreign Relations, 1865–1900.* Edited by Henry Steele Commager. New American Nation Series. New York: Harper and Row, 1976.

Canney, Donald L. *The Old Steam Navy: Frigates, Sloops, and Gunboats, 1815–1885.* 2 vols. Annapolis: Naval Institute Press, 1990.

Carrison, Capt. Daniel J. *The Navy from Wood to Steel, 1860–1890.* New York: Franklin Watts, 1965.

Carter, Samuel. *The Incredible Great White Fleet.* New York: Crowell-Collier, 1970.

Chadwick, Rear Adm. French Ensor. *The Relations of the United States and Spain: Diplomacy.* New York: Charles Scribner's Sons, 1909.

Chandler, Alfred. *The Visible Hand: The Managerial Revolution in American Business.* 16th ed. Cambridge: Belknap Press of Harvard University Press, 1977.

Chernow, Ron. *The House of Morgan: An American Banking Dynasty and the Rise of Modern Finance.* New York: Atlantic Monthly Press, 1990.

Chisholm, Donald. *Waiting for Dead Men's Shoes: Origins and Development of the U.S. Navy's Officer Personnel System, 1793–1941.* Stanford, CA: Stanford University Press, 2001.

Chua, Amy. *Day of Empire: How Hyperpowers Rise to Global Dominance—and Why They Fall.* New York: Doubleday, 2007.

Cogar, W. B. *Dictionary of Admirals of the U.S. Navy: 1862–1900.* Annapolis: Naval Institute Press, 1989.

Coletta, Paolo. *Admiral Bradley A. Fiske and the American Navy.* Lawrence: University Press of Kansas, 1979.

————. *French Ensor Chadwick: Scholarly Warrior.* Lanham, MD: University Press of America, 1980.

————. *A Selected and Annotated Bibliography of American Naval History.* Lanham, MD: University Press of America, 1988.

————. *A Survey of U.S. Naval Affairs, 1863–1917.* Lanham, MD: University Press of America, 1987.

————. *A Survey of U.S. Naval Affairs, 1865–1917.* Lanham, MD: University Press of America, 1987.

Collingwood, R. G., and J. Dussen. *The Idea of History: With Lectures 1926–1928.* New York: Oxford University Press, 1994.

Cooling, B. F. *Benjamin Franklin Tracy: Father of the Modern American Fighting Navy.* Hamden, CT: Archon Books, 1973.

————. *Gray Steel and Blue Water Navy.* Hamden, CT: Archon Books, 1979.

Cooper, Jerry M. *The Army and Civil Disorder: Federal Military Intervention in Labor Disputes, 1877–1900.* Edited by Thomas E. Griess and Jay Luvaas. Contributions in Military History. Westport, CT: Greenwood Press, 1980.

Corbett, Julian S. *Some Principles of Maritime Strategy.* London: Longmans, Green, 1911.

Cornebise, Alfred Emile. *Ranks and Columns: Armed Forces Newspapers in American Wars.* Contributions to the Study of Mass Media and Communications. Westport, CT: Greenwood Press, 1993.

Cornish, Dudley Taylor, and Virginia Jeans Laas. *Lincoln's Lee: The Life of Samuel Phillips Lee, United States Navy, 1812–1897.* Lawrence: University Press of Kansas, 1986.

Crichfield, G. W. *American Supremacy: The Rise and Progress of the Latin American Republics and Their Relations to the United States under the Monroe Doctrine.* New York: Brentano's, 1908.

Davis, George T. *A Navy Second to None: The Development of Modern American Naval Policy.* New York: Harcourt, Brace, 1940.

Doenecke, Justus D. *The Presidencies of James A. Garfield and Chester A. Arthur.* Edited by Donald R. McCoy, Clifford S. Griffin, and Homer E. Socolofsky. American Presidency Series. Lawrence: Regents Press of Kansas, 1981.

Dorwart, Jeffery M. *The Office of Naval Intelligence: The Birth of America's First Intelligence Agency, 1865–1918.* Annapolis, MD: Naval Institute Press, 1979.

Edgerton, Robert B. *Remember the Maine, to Hell with Spain: America's 1898 Adventure in Imperialism.* Lewiston, NY: Edwin Mellen Press, 2004.

Edwards, E. M. H. *Commander William Barker Cushing of the United States Navy.* New York: F. Tennyson Neely, 1898.

Elias, Norbert. *The Genesis of the Naval Profession.* Dublin: University College Dublin Press, 2007.

Evans, David Christian. *Building the Steam Navy: Dockyards, Technology, and the Creation of the Victorian Battle Fleet, 1830–1906.* Edited by English Heritage. London: Conway Maritime Press, 2004.

Evans, Rear Adm. Robley. *A Sailor's Log: Recollections of Forty Years of Naval Life.* New York: D. Appleton, 1901.

Fatherley, John A. *In the Vortex: Charles E. Clark, U.S.N.* West Springfield, MA: J. A. Fatherley, 2001.

Faust, D. G. *This Republic of Suffering: Death and the American Civil War.* New York: Vintage Books, 2009.

Felker, C. C. *Testing American Sea Power: U.S. Navy Strategic Exercises, 1923–1940.* College Station: Texas A&M University Press, 2006.

Ferguson, Niall. *Empire: The Rise and Demise of the British World Order and the Lessons for Global Power.* New York: Basic Books, 2003.

Feuer, A. B. *The Spanish-American War at Sea: Naval Action in the Atlantic.* Westport, CT: Praeger, 1995.

Fiske, Bradley A. *The Art of Fighting: Its Evolution and Progress with Illustrations from Campaigns of Great Commanders.* New York: Century, 1920.

———. *From Midshipman to Rear-Admiral:* New York: Century, 1919.

———. *The Navy as a Fighting Machine.* Edited by John B. Hattendorf and Wayne P. Hughes Jr. Classics of Seapower Series. Annapolis: Naval Institute Press, 1988.

Foner, Philip Sheldon. *The Great Labor Uprising of 1877.* New York: Monad Press, 1977.

Fremantle, E. R. *The Navy as I Have Known It: 1849–1899.* New York: Cassell, 1904.

Friedman, Norman. *U.S. Battleships: An Illustrated Design History.* Annapolis: Naval Institute Press, 1985.

———. *U.S. Cruisers: An Illustrated Design History.* Annapolis: Naval Institute Press, 1984.

Gleaves, Albert. *Life and Letters of Rear Admiral Stephen B. Luce, U.S. Navy, Founder of the Naval War College.* New York: G. P. Putnam's Sons, 1925.

Goldrick, James. *No Easy Answers: The Development of the Navies of India, Pakistan, Bangladesh and Sri Lanka, 1945–1996.* Edited by Royal Australian Navy Maritime Studies Program. Papers in Australian Maritime Affairs. Hartford, WI: Spantech and Lancer, 1997.

Gordon, Andrew. *The Rules of the Game: Jutland and British Naval Command.* London: John Murray Publishers, 2005.

Gough, Barry M. *The Royal Navy and the Northwest Coast of North America, 1810–1914: A Study of British Maritime Ascendancy.* Vancouver: University of British Columbia Press, 1971.

Grandin, Greg. *Empire's Workshop: Latin America, the United States, and the Rise of the New Imperialism.* Edited by Tom and Fraser Engelhardt, Steve, The American Empire Project. New York: Metropolitan Books/Henry Holt, 2007.

Gray, Edwin. *Nineteenth Century Torpedoes and Their Inventors.* Annapolis: Naval Institute Press, 2004.

Grenville, John A. S., and George Berkeley Young. *Politics, Strategy, and American Diplomacy: Studies in Foreign Policy, 1873–1917.* New Haven, CT: Yale University Press, 1966.

Haas, J. M. *A Management Odyssey: The Royal Dockyards, 1714–1914.* Lanham, MD: University Press of America, 1994.

Hackemer, Kurt. *The U.S. Navy and the Origins of the Military-Industrial Complex, 1847–1883.* Annapolis: Naval Institute Press, 2001.

Hagen, Kenneth J. *American Gunboat Diplomacy and the Old Navy, 1877–1889.* Contributions in Military History. Westport, CT: Greenwood Press, 1973.

————. *This People's Navy.* New York: Macmillan, 1991.

Harrod, Frederick S. *Manning the New Navy: The Development of a Modern Naval Enlisted Force, 1899–1940.* Edited by Jon L. Wakelyn. Contributions in American History. Westport, CT: Greenwood Press, 1978.

Hart, Edward H. *Squadron Evolutions: As Illustrated by the Combined Military and Naval Operations at Newport, R.I., November 1887, Rear Admiral S. B. Luce.* New York: E. H. Hart, 1887.

Hart, Robert A. *The Great White Fleet: Its Voyage around the World.* Boston: Little, Brown, 1965.

Haugen, Douglas Mark, *How Nineteenth-Century Naval Theorists Created America's Twentieth-Century Imperialist Policy: Military Strategy Shapes Foreign Policy,* Lewiston, NY: Edwin Mellen Press, 2010.

Healy, David. *James G. Blaine and Latin America.* Columbia: University of Missouri Press, 2001.

Hendrix, Henry. *Theodore Roosevelt's Naval Diplomacy: The U.S. Navy and the Birth of the American Century.* Annapolis: Naval Institute Press, 2009.

Herrick, Walter R. *The American Naval Revolution.* Baton Rouge: Louisiana State University Press, 1966.

Herring, George C. *From Colony to Superpower: U.S. Foreign Relations since 1776.* Edited by David M. Kennedy. Oxford History of the United States. New York: Oxford University Press, 2008.

Herwig, Holger H. *"Luxury" Fleet: The Imperial German Navy, 1888–1918.* Atlantic Highlands, NJ: Ashfield Press, 1987.

Hill, Richard. *War at Sea in the Ironclad Age.* Edited by John Keegan. Cassell's History of Warfare. London: Cassell, 2000.

Hofstadter, Richard. *The Age of Reform: From Bryan to FDR.* New York: Random House, 1955.

Hone, Thomas C., and Trent Hone. *Battle Line: The United States Navy, 1919–1939.* Annapolis: Naval Institute Press, 2006.

Hughes, Wayne P. *Fleet Tactics and Coastal Combat.* 2nd ed. Annapolis: Naval Institute Press, 2000.

Huntington, Samuel. *The Soldier and the State: The Theory and Politics of Civil-Military Relations.* 12th ed. Cambridge: Belknap Press of Harvard University Press, 1995.

Janowitz, Morris. *The Professional Soldier: A Social and Political Portrait.* New York: Free Press, 1971.

Jay, J. *The Fisheries Dispute: A Suggestion for Its Adjustment by Abrogating the Convention of 1818, and Resting on the Rights and Liberties Defined in the Treaty of 1783; a Letter to the Honourable William M. Evarts, of the United States Senate.* New York: Dodd, Mead, 1887.

Jones, Jerry W. *U.S. Battleship Operations in World War I.* Annapolis: Naval Institute Press, 1998.

Karsten, Peter. *The Naval Aristocracy: The Golden Age of Annapolis and the Emergence of Modern American Navalism.* Annapolis: Naval Institute Press, 2008.

Kennedy, Paul. *The Rise and Fall of British Naval Mastery.* Malabar, FL: Robert E. Krieger, 1982.

————. *The Rise and Fall of the Great Powers: Economic Change and Military Conflict from 1500 to 2000.* New York: Random House, 1987.

King, R. W. *Naval Engineering and American Sea Power.* Baltimore: Nautical & Aviation Publishing, 1989.

Kipling, R. *A Fleet in Being: Notes of Two Trips with the Channel Squadron.* London: B. Tauchnitz, 1899.

Kloppenberg, James. *Uncertain Victory: Social Democracy and Progressivism in European and American Thought, 1870–1920.* New York: Oxford University Press, 1986.

Kolko, Gabriel. *The Triumph of Conservatism: A Reinterpretation of American History, 1900–1916.* New York: Free Press, 1963.

Kramer, Paul A. *The Blood of Government: Race, Empire, the United States, and the Philippines.* Chapel Hill: University of North Carolina Press, 2006.

LaFeber, Walter. *The New Empire: An Interpretation of American Expansion, 1860–1898.* 4th ed. Ithaca, NY: Cornell University Press, 1963.

———. *The Panama Canal: The Crisis in Historical Perspective.* New York: Oxford University Press, 1989.

Lambert, Nicholas. *Sir John Fisher's Naval Revolution.* Edited by William N Still, Studies in Maritime History. Columbia: University of South Carolina Press, 2002.

Lamoreaux, Naomi. *The Great Merger Movement in American Business, 1895–1904.* Cambridge: Cambridge University Press, 1985.

Larson, Magali Sarfatti. *The Rise of Professionalism: A Sociological Analysis.* Berkeley and Los Angeles: University of California Press, 1977.

Lears, Jackson. *Rebirth of a Nation: The Making of Modern America, 1877–1920.* New York: Harper Collins, 2009.

Leeke, Jim. *Manila and Santiago: The New Steel Navy in the Spanish-American War.* Annapolis: Naval Institute Press, 2009.

Lens, Sidney. *The Forging of the American Empire: From the Revolution to Vietnam—a History of U.S. Imperialism.* London: Pluto Press, 2003.

Lewis, Charles Lee. *David Glasgow Farragut: Our First Admiral.* Annapolis: Naval Institute Press, 1943.

Lindsay-Poland, John. *Emperors in the Jungle: The Hidden History of the U.S. in Panama.* Edited by Gilbert M. Joseph and Emily S. Rosenberg. American Encounters / Global Interactions. Durham, NC: Duke University Press, 2003.

Logan, Rayford W. *The Diplomatic Relations of the United States with Haiti, 1776–1891.* Chapel Hill: University of North Carolina Press, 1941.

Long, John D. *The New American Navy.* 2 vols. New York: Outlook, 1903.

Love, Robert W., Jr. *History of the U.S. Navy.* Vol. 1, *1775–1941.* Harrisburg, PA: Stackpole Books, 1992.

Maclay, Edgar Stanton. *Reminiscences of the Old Navy: From the Journal and Private Papers of Captain Edward Trenchard, and Rear-Admiral Stephen Decatur Trenchard.* New York: G. P. Putnam's, 1898.

Mahan, Alfred T. *From Sail to Steam: Recollections of Naval Life.* New York: Harper and Brothers, 1907.

———. *The Influence of Sea Power Upon History.* 12th ed. Boston: Little, Brown, 1918.

———. *Naval Strategy, Compared and Contrasted with the Principles and Practice of Military Operations on Land.* Boston: Little, Brown, 1919.

Maier, Charles S. *Among Empires: American Ascendancy and Its Predecessors.* Cambridge: Harvard University Press, 2006.

Marder, Arthur J. *The Anatomy of British Sea Power: A History of British Naval Policy in the Pre-Dreadnaught Era, 1880–1905.* New York: Octagon Books, 1940.

———. *From the Dreadnaught to Scapa Flow: The Royal Navy in the Fisher Era, 1904–1919.* London: Oxford University Press, 1970.

May, Ernest R. *American Imperialism: A Speculative Essay.* New York: Atheneum, 1968.

McBride, William M. *Technological Change and the United States Navy, 1865–1945.* Edited by Merritt Roe Smith. Johns Hopkins Studies in the History of Technology. Baltimore: Johns Hopkins University Press, 2000.

McFeely, William S. *Grant: A Biography.* Paperback ed. New York: W. W. Norton, 1981.

McGerr, Michael. *A Fierce Discontent: The Rise and Fall of the Progressive Movement in America, 1870–1920.* New York: Oxford University Press, 2003.

McGraw, Thomas. *Prophets of Regulation.* Cambridge: Belknap Press of Harvard University Press, 1984.

McManemin, John A. *The Barrons: Four Commodores from Virginia.* Spring Lake, NJ: Ho Ho Kus, 1998.

Mead, Walter Russell. *God and Gold: Britain, America, and the Making of the Modern World.* New York: Alfred A. Knopf, 2007.

Military Historical Society of Massachusetts. *Naval Actions and History, 1799–1898.* Boston: Military Historical Society of Massachusetts, 1902.

Miller, Nathan. *The U.S. Navy: A History.* 3rd ed. Annapolis: Naval Institute Press, 1997.

Millett, Allan R. *The General: Robert L. Bullard and Officership in the United States Army, 1881–1925.* Westport, CT: Greenwood Press, 1975.

———. *Semper Fidelis: The History of the United States Marine Corps.* Edited by Louis Morton. Free Press Paperback Edition. Macmillan Wars of the United States. New York: Macmillan, 1980.

Millington, Herbert. *American Diplomacy and the War of the Pacific.* Edited by Faculty of Political Science of Columbia University. Studies in History, Economics, and Public Law. New York: Columbia University Press, 1948.

Mitchell, Donald W. *History of the Modern American Navy, from 1883 through Pearl Harbor.* New York: Alfred A. Knopf, 1946.

Montague, Ludwell Lee. *Haiti and the United States 1714–1938.* Durham, NC: Duke University Press, 1940.

Morgan, James Morris, and John Phillip Marquand. *Prince and Boatswain: Sea Tales from the Recollection of Rear Admiral Charles E. Clark.* Greenfield, MA: E. A. Hall, 1915.

Murphy, Gretchen. *Hemispheric Imaginings: The Monroe Doctrine and Narratives of U.S. Empire.* Durham: University of North Carolina Press, 2005.

The Nation. Vol. 60. New York: Nation Company, 1895.

Nenninger, Timothy K. *The Leavenworth Schools and the Old Army: Education, Professionalism, and the Officer Corps of the United States Army, 1881–1918.* Edited by Thomas E. Griess and Jay Luvaas. Contributions in Military History. Westport, CT: Greenwood Press, 1978.

Nevins, Allan. *Hamilton Fish: The Inner History of the Grant Administration*. 2nd ed. 2 vols. Vol. I, *American Classics*. New York: Frederick Ungar, 1967.

Nugent, Walter. *Habits of Empire: A History of American Expansionism*. New York: Alfred A. Knopf, 2008.

Otis, J. *The Boys of '98*: D. Estes: Howard University, 1898.

Owens, W. A. *High Seas: The Naval Passage to an Uncharted World*. Annapolis: Naval Institute Press, 1995.

Palmer, Fanny Ashurst. *The Loss of the Huron: A Vindication of Lieutenant Lambert G. Palmer*. Newport, RI: Davis and Pitman, 1883.

Palmer, Michael A. *Command at Sea: Naval Command and Control since the Sixteenth Century*. Cambridge: Harvard University Press, 2005.

Parker, Foxhall A. *Fleet Tactics under Steam*. New York: D. Van Nostrand, 1870.

———. *The Naval Howitzer Ashore*. New York: D. Van Nostrand, 1866.

———. *Squadron Tactics under Steam*. New York: D. Van Nostrand, 1864.

Parker, Geoffrey. *The Army of Flanders and the Spanish Road, 1567–1659*. Cambridge Studies in Early Modern History. Series editors H. G. Koenigsberger and J. H. Elliott. New York: Cambridge University Press, 1972.

Parkinson, R. *The Late Victorian Navy: The Pre-Dreadnought Era and the Origins of the First World War*. Woodbridge, UK: Boydell Press, 2008.

Paullin, Charles Oscar. *Paullin's History of Naval Administration, 1775–1911*. Annapolis: Naval Institute Press, 1968.

Peltier, Eugene. *The Bureau of Yards and Docks of the Navy and the Civil Engineer Corps*. New York: Newcomen Society in America, 1961.

Perez, Louis A., Jr. *On Becoming Cuban: Identity, Nationality, and Culture*. Chapel Hill: University of North Carolina Press, 1999.

———. *The War of 1898: The United States and Cuba in History and Historiography*. Chapel Hill: University of North Carolina Press, 1998.

Plesur, Milton. *America's Outward Thrust: Approaches to Foreign Affairs, 1865–1890*. DeKalb: Northern Illinois University Press, 1971.

———. *Creating an American Empire, 1865–1914*. Edited by A. S. Eisenstadt. Major Issues in American History. New York: Pitman, 1971.

Potter, E. B. *Sea Power: A Naval History*. 2nd ed. Annapolis, MD: U.S. Naval Institute, 1981.

Pratt, F. *Preble's Boys: Commodore Preble and the Birth of American Sea Power*. New York: Sloane, 1950.

Preston, Anthony, and John Major. *Send a Gunboat: The Victorian Navy and Supremacy at Sea, 1854–1904*. London: Conway Maritime, 2007.

Preston, Richard A. *Perspectives in the History of Military Education and Professionalism*. The Harmon Memorial Lectures in Military History. Colorado Springs: United States Air Force Academy, 1980.

Pugh, Philip. *The Cost of Seapower: The Influence of Money on Naval Affairs from 1815 to the Present Day*. London: Conway Maritime Press, 1986.

Reason, Adm. J. Paul. *Sailing New Seas*. Edited by Patricia A. Goodrich. The Newport Papers. Newport, RI: Naval War College Press, 1998.

Reckner, James. *Teddy Roosevelt's Great White Fleet*. Annapolis: Naval Institute Press, 1988.

Reynolds, Clark G. *Famous American Admirals*. First Naval Institute Press Edition. Annapolis: Naval Institute Press, 2002.

Richard, Alfred Charles, Jr. *The Panama Canal in American National Consciousness, 1870–1990*. New York: Garland Publishing, 1990.

Rodger, N. A. M. *The Wooden World: An Anatomy of the Georgian Navy*. Annapolis: Naval Institute Press, 1986.

Rodgers, Daniel T. *Atlantic Crossings: Social Politics in a Progressive Age*. Cambridge: Belknap Press of Harvard University Press, 1998.

Rodman, Rear Adm. Hugh, USN. *Yarns of a Kentucky Admiral*. Indianapolis: Bobbs-Merrill, 1928.

Roosevelt, R. *The Works of Theodore Roosevelt: American Ideals, with a Biographical Sketch by F. V. Greene: Administration*. New York: P. F. Collier, 1897.

Ropp, Theodore. *The Development of a Modern Navy: French Naval Policy, 1871–1904*. Edited by Stephen S. Roberts. Annapolis: Naval Institute Press, 1987.

Roske, Ralph J., and Charles Van Doren. *Lincoln's Commando: The Biography of Cdr. William B. Cushing, USN*. Bluejacket Books edition. Annapolis: Naval Institute Press, 1995.

Roskill, S. W. *The Strategy of Seapower: Its Development and Application*. Glasgow: Collins, 1961.

Royal United Service Institution. *Journal of the Royal United Service Institution*. Mitchell, 1869.

Rüger, Jan. *The Great Naval Game: Britain and Germany in the Age of Empire*. Edited by Jay Winter. Studies in the Social and Cultural History of Modern Warfare. New York: Cambridge University Press, 2007.

Ruether, Rosemary Radford. *America, Amerikkka: Elect Nation and Imperial Violence*. Sheffield, UK: Equinox Publishing, 2007.

Rydell, R. W. *All the World's a Fair: Visions of Empire at American International Expositions, 1876–1916*. Chicago: University of Chicago Press, 1987.

———. *World of Fairs: The Century-of-Progress Expositions*. Chicago: University of Chicago Press, 1993.

Sater, William F. *Andean Tragedy: Fighting the War of the Pacific, 1879–1884*. Edited by Peter Maslowski, David Graff, and Reina Pennington. Studies in War, Society, and the Military. Lincoln: University of Nebraska Press, 2007.

Schneller, Robert J., Jr. *Cushing: Civil War Seal*. Edited by Dennis E. Showalter. Military Profiles. Washington, DC: Brassey's, 2004.

Schrijvers, Peter. *The GI War against Japan: American Soldiers in Asia and the Pacific During World War Ii*. New York: New York University Press, 2002.

Segal, David R., and Joseph J. Lengermann. "Professional and Institutional Considerations." In *Combat Effectiveness: Cohesion, Stress, and the Volunteer Military*, edited by Sam C. Sarkesian, 154–84. Beverly Hills, CA: Sage Publications, 1980.

Seymour, E. H. *My Naval Career and Travels*. London: Smith, Elder, 1911.

Shulimson, Jack. *The Marine Corps Search for a Mission, 1880–1898*. Edited by Theodore A. Wilson. Modern War Studies. Lawrence: University Press of Kansas, 1993.

Shulman, Mark R. *Navalism and the Emergence of American Sea Power, 1882–1893*. Annapolis: Naval Institute Press, 1995.

Simpson, Edward. *Yarnlets: The Human Side of the Navy*. New York: Putnam, 1934.

Skowronek, Stephen. *Building a New American State: The Expansion of National Administrative Capacities, 1877–1920.* Cambridge: Cambridge University Press, 1982.

Soley, James Russell. *Admiral Porter.* New York: D. Appleton, 1903.

——. *Historical Sketch of the United States Naval Academy.* Washington, DC: GPO, 1876.

Spector, Ronald. *At War, at Sea: Sailors and Naval Warfare in the Twentieth Century.* New York: Penguin Group, 2001.

——. *Professors of War: The Naval War College and the Development of the Naval Profession.* Edited by B. M. Simpson III. Naval War College Historical Monograph Series. Newport, RI: Naval War College Press, 1977.

Sprout, Harold, and Margaret Sprout. *The Rise of American Naval Power, 1776–1918.* Edited by Jack Sweetman. Classics of Naval Literature. Annapolis: Naval Institute Press, 1990.

Still, William N. *American Sea Power in the Old World: The United States Navy in European and Near Eastern Waters, 1865–1917.* Westport, CT: Greenwood Press, 1980.

——. *Crisis at Sea: The United States Navy in European Waters in World War I.* Foreword by James C. Smith and Gene A. Bradford. New Perspectives on Maritime History and Nautical Archaeology. Gainesville: University Press of Florida, 2006.

Stokesbury, James L. *Navy and Empire.* New York: William Morrow, 1983.

Sumida, Jon. *In Defense of Naval Supremacy: Finance, Technology, and British Naval Policy, 1889–1914.* London: Routledge, 1993.

——. *Inventing Grand Strategy and Teaching Command: The Classic Works of Alfred Thayer Mahan Reconsidered.* Washington, DC: Woodrow Wilson Center Press, 1997.

Sumida, Jon T. and David A. Rosenberg, "Machines, Men, Manufacturing, Management, and Money: The Study of Navies as Complex Organizations and the Transformation of Twentieth Century Naval History," in *Doing Naval History: Essays Towards Improvement,* ed. John B. Hattendorf, Naval War College Historical Monograph Series (Newport, RI: Naval War College Press, 1995)

Swann, Leonard Alexander, Jr. *John Roach Maritime Entrepreneur: The Years as Naval Contractor, 1862–1886.* Annapolis: Naval Institute Press, 1965.

Symonds, Craig L. *The Naval Institute Historical Atlas of the U.S. Navy.* Annapolis: Naval Institute Press, 1995.

Taylor, Bruce. *The Battlecruiser Hood: An Illustrated Biography, 1916–1941.* Annapolis: Naval Institute Press, 2008.

Teitler, G. *The Genesis of the Professional Officer Corps.* Translated by C. N. Ter Heide-Lopy. Edited by Morris Janowitz and Charles C. Moskos Jr. Inter-University Seminar on Armed Forces and Society. Beverly Hills, CA: Sage Publications, 1977.

Tenner, Edward. *Why Things Bite Back: Technology and the Revenge of Unintended Consequences.* New York: Alfred A. Knopf, 1996.

Trachtenberg, Alan. *The Incorporation of America: Culture and Society in the Gilded Age.* Edited by Eric Foner. American Century Series. New York: Hill and Wang, 1982.

Tucker, Spencer C. *Handbook of 19th Century Naval Warfare.* Annapolis: Naval Institute Press, 2000.

Uhlig, Frank, Jr. *How Navies Fight: The U.S. Navy and Its Allies.* Annapolis: Naval Institute Press, 1994.

U.S. House Committee on Foreign Affairs, F. D. Pavey, W. N. Cromwell, and P. Bunau-Varilla. *The Story of Panama: Hearings on the Rainey Resolution before the Committee on Foreign Affairs of the House of Representatives.* Washington, DC: GPO, 1913.

Van Creveld, Martin. *The Training of Officers: From Military Professionalism to Irrelevance.* New York: Free Press, 1990.

Very, Lt. E. W., USN. *Report on Torpedo-Boats for Coast Defense.* Washington, DC: GPO, 1884.

Watt, Donald Cameron. *Succeeding John Bull: America in Britain's Place, 1900–1974: A Study of the Anglo-American Relationship and World Politics in the Context of British and American Foreign-Policy-Making in the Twentieth Century.* New York: Cambridge University Press, 1984.

Wiebe, Robert H. *The Search for Order, 1877–1920.* 1st ed. Edited by David Donald. The Making of America. New York: Hill and Wang, 1967.

Williams, William Appleman. *Empire as a Way of Life: An Essay on the Causes and Character of America's Present Predicament Along with a Few Thoughts About an Alternative.* New York: Oxford University Press, 1980.

———. *The Roots of the Modern American Empire: A Study of the Growth and Shaping of Social Consciousness in a Marketplace Society.* New York: Random House, 1969.

Willis, Sam. *Fighting at Sea in the Eighteenth Century: The Art of Sailing Warfare.* Woodbridge, UK: Boydell Press, 2008.

Winklareth, R. J. *Naval Shipbuilders of the World: From the Age of Sail to the Present Day.* London: Chatham, 2000.

Winkler, Jonathan Reed. *Nexus: Strategic Communications and American Security in World War I.* Cambridge: Harvard University Press, 2008.

Zunz, Oliver. *Making America Corporate, 1870–1920.* Chicago: University of Chicago Press, 1990.

Conference Papers and Proceedings

Dorwart, Jeffery M. "Naval Attaches, Intelligence Officers, and the Rise of the 'New American Navy,' 1882–1914." In *Third United States Naval Academy History Symposium,* edited by Robert W. Love Jr., 260–69. New York: Garland Publishing, 1980.

Hagen, Kenneth J., B. Franklin Cooling, Jacob W. Kipp, Bruce Swanson, and A. Michal McMahon. "Naval Technology and Social Modernization in the Nineteenth Century." Paper presented at the Society for the History of Technology, Chicago, 1974.

McBride, William M. "New Interpretations in Naval History." Paper presented at the Thirteenth Naval History Symposium, Annapolis, MD, 1997.

Roberts, Stephen S. "An Indicator of Informal Empire: Patterns of U.S. Navy Cruising on Overseas Stations, 1869–1897." In *Fourth Naval History Symposium,* edited by Craig L. Symonds, 253–267. Annapolis: Naval Institute Press, 1979.

Edited Books

Bradford, James C., ed. *Admirals of the New Steel Navy: Makers of the Modern American Naval Tradition, 1880–1930.* Annapolis: Naval Institute Press, 1990.

———. *Captains of the Old Steam Navy: Makers of the American Naval Tradition, 1840–1880.* Makers of the American Naval Tradition. Annapolis: Naval Institute Press, 1986.

———. *Crucible of Empire: The Spanish-American War and Its Aftermath.* Annapolis: Naval Institute Press, 1993.

Calhoun, Craig, Frederick Cooper, and Kevin W. Moore, eds. *Lessons of Empire: Imperial Histories and American Power.* New York: New Press, 2006.

Cassard, Chaplain William G., USN, ed. *Battleship* Indiana *and Her Part in the Spanish-American War.* New York: Everett B. Mero, 1898.

Chesneau, Roger, and Eugene M. Kolesnik, eds. *Conway's All the World's Fighting Ships, 1860–1905.* London: Conway Maritime Press, 1979.

Coletta, Paolo, ed. *American Secretaries of the Navy.* Vol. 1, *1775–1913.* Annapolis: Naval Institute Press, 1980.

Esherick, Joseph, Hasan Kayali, and Eric Van Young, eds. *Empire to Nation: Historical Perspectives on the Making of the Modern World.* Lanham, MD: Rowman and Littlefield, 2006.

Goode, W. A. M., ed. *With Sampson through the War: Being an Account of the Naval Operations of the North Atlantic Squadron During the Spanish-American War of 1898.* New York: Doubleday and McClure, 1899.

Hagan, Kenneth J., ed. *In Peace and War: Interpretations of American Naval History.* 30th Anniversary Edition. Westport, CT: Praeger Security International, 2008.

Hattendorf, John B., ed. *Doing Naval History: Essays Towards Improvement.* Naval War College Historical Monograph Series. Newport, RI: Naval War College Press, 1995.

Hattendorf, John B., and John D. Hayes, eds. *The Writings of Stephen B. Luce.* Naval War College Historical Monograph Series. Newport, RI: Naval War College Press, 1975.

Hattendorf, John B., and Robert S. Jordan, eds. *Maritime Strategy and the Balance of Seapower: Britain and America in the Twentieth Century.* New York: St. Martin's Press.

Karsten, Peter, ed. *The Military in America: From the Colonial Era to the Present.* New York: Free Press, 1986.

Lynn, John A., ed. *Tools of War: Instruments, Ideas, and Institutions of Warfare, 1445–1871.* Chicago: University of Illinois Press, 1990.

Maguire, Doris D. *French Ensor Chadwick: Selected Letters and Papers.* Washington, DC: University Press of America, 1981.

Margiotta, Franklin D., ed. *The Changing World of the American Military.* Westview Special Studies in Military Affairs. Boulder, CO: Westview Press, 1978.

Marolda, Edward J., ed. *Theodore Roosevelt, the U.S. Navy, and the Spanish-American War.* Introduction by Douglas Brinkley. Franklin and Eleanor Roosevelt Institute on Diplomatic and Economic History. New York: Palgrave, 2001.

Merli, Frank J., and Theodore A. Wilson, eds. *Makers of American Diplomacy: From Benjamin Franklin to Alfred Thayer Mahan.* New York: Charles Scribner's Sons, 1974.

Sarkesian, Sam C., ed. *Combat Effectiveness: Cohesion, Stress, and the Volunteer Military.* 9 vols. Sage Research Progress Series on War, Revolution, and Peacekeeping. Beverly Hills, CA: Sage Publications, 1980.

Schirmer, Daniel B., and Stephen Rosskamm Shalom, eds. *The Philippines Reader: A History of Colonialism, Neocolonialism, Dictatorship, and Resistance.* Boston: South End Press, 1897.

Smith, Angel, and Emma Davila-Cox, eds. *The Crisis of 1898: Colonial Redistribution and Nationalist Mobilization.* New York: Palgrave Macmillan, 1999.

Stoler, Ann Laura, Carole McGranahan, and Peter C. Perdue, eds. *Imperial Formations.* Santa Fe, NM: School for Advanced Research Press, 2007.

Sumida, Jon T., and David A. Rosenberg. "Machines, Men, Manu- facturing, Management, and Money: The Study of Navies as Complex Organizations and the Transformation of Twentieth Naval History." In *Doing Naval History: Essays towards Improvement,* edited by John B. Hattendorf. Naval War College Historical Monograph Series. Newport, RI: Naval War College Press, 1995.

Till, Geoffrey, ed. *The Development of British Naval Thinking: Essays in Memory of Bryan Mclaren Ranft.* Cass Series: Naval Policy and History. Abingdon, UK: Routledge, 2006.

Trachtenberg, Alan, ed. *Democratic Vistas, 1860–1880.* American Culture Series. New York: George Braziller, 1970.

Turner, Frederick Jackson. "The Significance of the Frontier in American History." In *History, Frontier, and Section: Three Essays by Frederick Jackson Turner,* by Frederick Jackson Turner. Albuquerque: University of New Mexico Press, 1993.

Government Documents

Chadwick, French Ensor. *Report on the Training Systems for the Navy and Mercantile Marine of England and on the Naval Training System of France Made to the Bureau of Equipment and Recruiting, U.S. Navy Department, September 1879.* 46th Cong., 2nd sess., 1880.

Department of the Navy, Annual Report of the Secretary of the Navy on the Operations of the Department, with Accompanying Documents for the Year 1866.

Letter from the Secretary of the Navy Reporting, in Answer to Senate Resolution of the 4th Instant, the Steps Taken by Him to Establish and Advanced Course of Instruction of Naval Officers at Coasters' Island, Rhode Island. 48th Cong., 2 sess., 1885.

The Parliamentary Debates (Authorized Edition). London: H.M. Stationery Office, 1872.

U.S. Department of the Navy. Office of Naval Intelligence. "The American Naval Planning Section London." Washington, DC: GPO, 1923.

U.S. Department of State. "Message of the President Relating to the Steamer Virginius, with the Accompanying Documents Transmitted to Congress, January 5, 1874." Edited by Department of State. Washington, DC: GPO, 1874.

U.S. Department of State. *Papers Relating to the Foreign Relations of the United States. 1885, 244-245.*

U.S. House of Representatives, Committee on Foreign Affairs, *The Story of Panama: Hearings on the Rainey Resolution before the Committee on Foreign Affairs of the House of Representatives* (Washington, DC: GPO, 1913).

U.S. House of Representatives. Condition of the Navy: Letter from the Secretary of the Navy, in Response to a Resolution from the House of Representatives Requesting the Views of the Minority of the Commission to Consider the Condition of the Navy. 47th Cong., 1st sess., 1882. Executive Document No. 30.

Articles

Abbott, Andrew. "Jurisdictional Conflicts: A New Approach to the Development of the Legal Profession." *American Bar Foundation Research Journal* 11, no. 2 (1986): 187–224.

Albert, Stuart, and David Whetten. "Organizational Identity." *Research in Organizational Behavior* 7 (1985): 263–295.

Allin, Lawrence Carroll. "The First Cubic War: The Virginius Affair." *American Neptune,* October 1978, 233–248.

Ammen, Daniel. "The Purposes of a Navy, and the Best Methods of Rendering It Efficient." U.S. Naval Institute *Proceedings* 5, no. 7 (1879): 119–132.

Angevine, Robert G. "The Rise and Fall of the Office of Naval Intelligence, 1882–1892: A Technological Perspective." *Journal of Military History* 62 (April 1998): 291–312.

Ashforth, Blake E., and Fred Mael. "Social Identity Theory and the Organization." *Academy of Management Review* 14, no. 1 (1989): 20–39.

Barnett, William P., and Glenn R. Carroll. "Modeling Internal Organizational Change." *Annual Review of Sociology* 21 (1995): 217–236.

Barr, E. L. "The Navy Signal System." U.S. Naval Institute *Proceedings* 39, no. 2 (1913): 585–587.

Bastert, Russell H. "Diplomatic Reversal: Frelinghuysen's Opposition to Blaine's Pan-American Policy in 1882." *Mississippi Valley Historical Review* 42, no. 4 (1956): 653–671.

———. "A New Approach to the Origins of Blaine's Pan-American Policy." *Hispanic American Historical Review* 39, no. 3 (1959): 375–412.

Baxter, James Phinney, III. "June Meeting: The British High Commissioners at Washington 1871." *Proceedings of the Massachusetts Historical Society* 65 (October 1832–May 1936), no. 3 (1936): 334–357.

Becker, Howard S., and James Carper. "The Elements of Identification with an Occupation." *American Sociological Review* 21, no. 3 (1956): 8.

Belknap, Charles. "The Naval Policy of the United States." U.S. Naval Institute *Proceedings* 6 no. 13 (April 1880): 380.

Belknap, R. R. "Cruising in Formation." U.S. Naval Institute *Proceedings* 38, no. 1 (1912): 195–210.

Bernotti, Lt. Romeo. "The Fundamentals of Naval Tactics." U.S. Naval Institute *Proceedings* 37, no. 3 (1911): 877–932.

Black, Jeremy. "Frontiers and Military History." *Journal of Military History* 72, no. 4 (2008): 1047–1059.

Brown, Andrew D., and Ken Starkey. "Organizational Identity and Learning: A Psychodynamic Perspective." *Academy of Management Review* 25, no. 1 (2000): 102–120.

Bryan, H. F., P. R. Alger, and B. A. Fiske. "Gun Distribution Aboard Modern Battleships, and Its Influence on Naval Tactics." U.S. Naval Institute *Proceedings* 33, no. 1 (1907): 205–238.

Calkins, Carlos Gilman. "Tradition and Progress in the Navy: A Review of Service Opinions, 1841–1901." U.S. Naval Institute *Proceedings* 39, no. 3 (1913): 1189–11216.

Chamberlain, Lt. Henry, RN. "The Inefficiency of Our Present System of Naval Education." *Colburn's United Service Magazine* 2, no. 11 (1889): 550–556.

———. "The Inefficiency of Our Present System of Naval Education, Part II." *Colburn's United Service Magazine* 2, no. 12 (1889): 696–709.

Clowes, William Laird. "Consideration on the Battleship in Action." U.S. Naval Institute *Proceedings* 20, no. 2 (1894): 293–299.

Collins, Lt. Frederick, USN. "Naval Affairs." U.S. Naval Institute *Proceedings* 5, no. 8 (1879): 159–179.

Crosley, W. S. "The Naval War College, the General Board, and the Office of Naval Intelligence." U.S. Naval Institute *Proceedings* 39, no. 3 (1913): 966–974.

Dawes, R. A. "Battle Tactics." U.S. Naval Institute *Proceedings* 41, no. 6 (1915): 1873–1195.

Dennett, Tyler. "Seward's Far Eastern Policy." *American Historical Review* 28, no. 1 (1922): 45–62.

"The Devastation." *Colburn's United Service Magazine* 136, no. 3 (1874): 427–439.

Dubassov, Captain, Imperial Russian Navy. "Torpedo Warfare." *Colburn's United Service Magazine* 2, no. 10 (1889): 395–416.

Dutton, Jane E., and Janet M. Dukerich. "Keeping an Eye on the Mirror: Image and Identity in Organizational Adaptation." *Academy of Management Journal* 34, no. 3 (1991): 517–554.

Ellicott, Capt. J. M. "Three Navy Cranks and What They Turned." U.S. Naval Institute *Proceedings* 50, no. 10 (1924): 1615–1628.

Epstein, Katherine C. "No One Can Afford to Say 'Damn the Torpedoes': Battle Tactics and U.S. Naval History Before WWI." *Journal of Military History* 77, no. 2 (April 2013): 491–520.

Fairbank, John K. "'American China Policy' to 1898: A Misperception." *Pacific Historical Review* 39, no. 4 (1970): 409–420.

A Flag Officer. "Fighting Power of Our Navy." *Colburn's United Service Magazine* 2, no. 10 (1889): 453–471.

———. "Our War Fleet and Its Guns." *Colburn's United Service Magazine* 2, no. 12 (1889): 732–740.

Gerstle, Gary. "The Protean Character of American Liberalism." *American Historical Review* 99, no. 4 (1994): 1043–1073.

Gioia, Dennis A., and James B. Thomas. "Identity, Image, and Issue Interpretation: Sensemaking During Strategic Change in Academia." *Administrative Science Quarterly* 41, no. 3 (1996): 370–403.

Gorringe, Henry H. "The Navy." *North American Review* 134, no. 306 (1882): 486–506.

Greene, S. Dana, ENS. "Electricity on Board Warships." U.S. Naval Institute *Proceedings* 15, no. 3 (1889): 471–489.

Gresle, Francois. "The 'Military Society': Its Future Seen through Professionalism." *Revue française de sociologie* 46, no. Supplement: An Annual English Selection (2005): 37–57.

Grieve, Andrew P. "The Professionalization of The 'Shoe Clerk.'" *Journal of the Royal Statistical Society. Series A (Statistics in Society)* 168, no. 4 (2005): 639–656.

Grusky, Oscar. "Leadership and Commitment in the Royal Navy." *Maritime Studies and Management* 1, no. 4 (1974): 223–231.

Herbert, Hilary A. "The Lesson of the Naval Review." *North American Review* 156, no. 439 (1893): 641–647.

Hunter, Mark C. "The U.S. Naval Academy and Its Summer Cruises: Professionalization in the Antebellum U.S. Navy, 1845–1861." *Journal of Military History* 70, no. 4 (2006): 963–994.

Ingersoll, R. E. "Organization of the Fleet for War." U.S. Naval Institute *Proceedings* 39, no. 4 (1913): 1379–1405.

Janowitz, Morris. "Changing Patterns of Organizational Authority: The Military Establishment." *Administrative Science Quarterly* 3, no. 4 (1959): 473–493.

Knox, Dudley W. "'Column' as a Battle Formation." U.S. Naval Institute *Proceedings* 39, no. 3 (1913): 949–963.

Luce, Rear Adm. Stephen B. "The Fleet." *North American Review* (October 1908): 564–576.

———. "The Manning of Our Navy and Mercantile Marine." U.S. Naval Institute *Proceedings* 1, no. 1 (1875): 17–38.

———. "The Navy and Its Needs." *North American Review* (April 1911): 494–507.

―――. "Our Future Navy." *North American Review* 149, no. 392 (1889): 54–65.

―――. "Our Future Navy." *U.S. Naval Institute Proceedings* 15, no. 4 (1889).

―――. "A Plea for an Engineer Corps in the Navy." *North American Review* (January 1906): 74–83.

MacDonald, Paul K. "Those Who Forget Historiography Are Doomed to Republish It: Empire, Imperialism and Contemporary Debates About American Power." *Review of International Studies* 35, no. 01 (2009): 45–67.

Martinez-Fernandez, Luis. "Caudillos, Annexationism, and the Rivalry between Empires in the Dominican Republic, 1844–1874." *Diplomatic History* 17, no. 4 (Fall 1993): 571–597.

Mason, Lt. T. B. M. "Two Lessons from the Future." U.S. Naval Institute *Proceedings* 1, no. 4 (1875).

May, Ernest R. "The Nature of Foreign Policy: The Calculated Versus the Axiomatic." *Daedalus* 91, no. 4 (1962): 653–667.

McCormick, Thomas. "Insular Imperialism and the Open Door: The China Market and the Open Door." *Pacific Historical Review* 32, no. 2 (1963): 155–169.

Morison, Elting E. "A Case Study of Innovation." *Engineering and Science Magazine* 13, no. 7 (April 1950): 5–11.

―――. "Inventing a Modern Navy." *American Heritage* 37, no. 4 (June–July 1986): 81–96.

"Naval and Military Intelligence: Strength of the Royal Navy in Commission." *Colburn's United Service Magazine* 1, no. 545 (1974): 531–552.

"Naval Brigades." *Colburn's United Service Magazine* 1, no. 547 (1874): 143–155.

"The Navy Estimates." *Colburn's United Service Magazine* 1, no. 546 (1874): 78–84.

"The Navy in 1874." *Colburn's United Service Magazine* 1, no. 542 (1874): 83–94.

Niblack, A. P. "Discussion of Prize Essay, 1895." U.S. Naval Institute *Proceedings* 21, no. 2 (1895): 271–274.

―――. "The Elements of Fleet Tactics." U.S. Naval Institute *Proceedings* 32, no. 2 (1906): 387–445.

―――. "The Tactics of Ships in the Line of Battle." U.S. Naval Institute *Proceedings* 22, no. 1 (1896): 1–54.

Parker, Foxhall A. "Our Fleet Maneuvers in the Bay of Florida, and the Navy of the Future." U.S. Naval Institute Proceedings 1, no. 8 (1874): 163–178.

Parsons, Capt. T. E., R.N. "In Paraguayan Waters, Part I." *Colburn's United Service Magazine* 1, no. 547 (1874): 198–204.

―――. "Notes from a Journal in Paraguayan Waters, Part II." *Colburn's United Service Magazine* 1, no. 551 (1874): 191–205.

Pearce, George F. "The United States Navy Comes to Pensacola." *Florida Historical Quarterly* 55, no. 1 (1976): 37–47.

Peltier, Eugene. *The Bureau of Yards and Docks of the Navy and the Civil Engineer Corps* (New York: Newcomen Society in America, 1961).

Rodgers, Daniel T. "In Search of Progressivism." *Reviews in American History* 10, no. 4 (1982): 113–132.

Rüger, Jan. "Nation, Empire, and Navy: Identity Politics in the United Kingdom, 1887–1914." *Past and Present* 185 (2004): 159–188.

Sampson, Capt. W. T. "The Naval Defense of the Coast." U.S. Naval Institute *Proceedings* 15, no. 2 (1889): 169–232.

Schroeder, Lt. Seaton. "The Type of (I) Armored Vessel, (II) Cruiser Best Suited to the Present Needs of the United States." U.S. Naval Institute *Proceedings* 7, no. 1 (1881).

Seager, Robert II. "Ten Years Before Mahan: The Unofficial Case for the New Navy, 1880–1890." *Mississippi Valley Historical Review* 40, no. 3 (December 1953): 491–512

Segal, David R. "Selective Promotion in Officer Cohorts." *Sociological Quarterly* 8, no. 2 (1967): 199–206.

Segal, David R., and Mady Wechsler Segal. "Change in Military Organization." *Annual Review of Sociology* 9 (1983): 151–170.

Sergent, Nathan E. "Suggestions in Favor of More Efficient Fleet Exercises." U.S. Naval Institute *Proceedings* 10, no. 2 (1884): 234–235.

"The Service Clubs: The Royal Naval Club, Portsmouth." *Illustrated Naval and Military Magazine: A Monthly Journal Devoted to All Subjects Connected with Her Majesty's Land and Sea Forces* 3 (1885): 50–51.

Shulimson, Jack. "Military Professionalism: The Case of the U.S. Marine Officer Corps, 1880–1898." *Journal of Military History* 60, no. 2 (1996): 231–242.

Sigsbee, Cdr. C. D. "Progressive Naval Seamanship." U.S. Naval Institute *Proceedings* 15, no. 1 (1889): 95–129.

Soley, James Russell. "On a Proposed Type of Cruiser for the United States Navy." U.S. Naval Institute *Proceedings* 4, no. 8 (1878): 127–140.

Staunton, S. A. "Squadron Drills at Sea." *Harper's Weekly,* 31 March 1895, 299–301.

Sumida, Jon. "Reimagining the History of Twentieth-Century Navies," Daniel Finamore, ed., *Maritime History as World History* (Gainesville: University of Florida Press, 2004).

Sweitzer, Vicki Baker. "Towards a Theory of Doctoral Student Professional Identity Development: A Developmental Networks Approach." *Journal of Higher Education* 80, no. 1 (2009): 1–33.

"Swift Unarmoured Cruisers." *Colburn's United Service Magazine* 1, no. 544 (1874): 317–330.

"Torpedoes and Sunken Mines." *Colburn's United Service Magazine* 1, no. 551 (1874).

Very, Lt. E. W. "The Type of (I) Armored Vessel, (Ii) Cruiser, Best Suited to the Present Needs of the United States." U.S. Naval Institute *Proceedings* 7, no. 15 (1881): 43–83.

Volkoff, Olga, Diane M. Strong, and Michael B. Elmes. "Technological Embeddedness and Organizational Change." *Organization Science* 18, no. 5 (2007): 832–848.

Volwiler, A. T. "Harrison, Blaine, and American Foreign Policy, 1889–1893." *Proceedings of the American Philosophical Society* 79, no. 4 (1938): 637–648.

Wainwright, Lt. Richard. "Naval Coast Signals." U.S. Naval Institute *Proceedings* 15, no. 1 (1889): 61–74.

———. "Tactical Problems in Naval Warfare." U.S. Naval Institute *Proceedings* 21, no. 2 (1895): 45.

Watson, Samuel. "Continuity in Civil-Military Relations and Expertise: The U.S. Army During the Decade before the Civil War." *Journal of Military History* 75, no. 1 (January 2011): 221–250.

Willis, Sam. "Fleet Performance and Capability in the Eighteenth-Century Royal Navy." *War in History* 11, no. 4 (2004): 373–392.

Wolters, Timothy S. "Recapitalizing the Fleet: A Material Analysis of Late-Nineteenth-Century U.S. Naval Power." *Technology and Culture* 52, no. 1 (2011): 24.

Yarnell, Lt. Cdr. Harry E. "The Greatest Need of the Atlantic Fleet." U.S. Naval Institute *Proceedings* 39, no. 1 (1913): 40.

Theses and Dissertations

Buhl, Lance. "The Smooth Water Navy: American Naval Policy and Politics, 1865–1876." Ph.D. diss., Harvard University, 1968.

Cooling, B. F. "Benjamin Franklin Tracy: Lawyer, Soldier, Secretary of the Navy." Ph.D. diss., University of Pennsylvania, 1969.

Crumley, Brian T. "The Evolution of the 'New' American Navy and the Naval Aspects of the Spanish-American War." Master's thesis, University of Louisville, 1997.

———. "The Naval Attaché System of the United States, 1882–1914." Ph.D. diss., Texas A&M University, 2002.

Daellenbach, Dennis A. "Senators, the Navy, and the Politics of American Expansionism, 1881–1890." Ph.D. diss., University of Kansas, 1982.

Davidson, Roger A., Jr. "Yankee Rivers, Rebel Shore: The Potomac Flotilla and Civil Insurrection in the Chesapeake Region." Ph.D. diss., Howard University, 2000.

Drake, Frederick Charles. "The Empire of the Seas: A Biography of Robert Wilson Shufeldt, USN." Ph.D. diss., Cornell University, 1970.

Etnyre, Robb P. "Naval Leadership and Society." Master's thesis, Naval Postgraduate School, 1997.

Evans, David Christian. "The Satsuma Faction and Professionalism in the Japanese Naval Officer Corps of the Meiji Period, 1868–1912." Ph.D. diss., Stanford University, 1978.

Gilliam, Bates McCluer. "The World of Captain Mahan." Ph.D. diss., Princeton University, 1961.

Havern, Christopher B. "A Gunnery Revolution Manqué: William S. Sims and the Adoption of Continuous-Aim in the United States Navy, 1989–1910." Master's thesis, University of Maryland, College Park, 1995.

Heitzmann, William Ray. "The United States Naval Institute's Contribution to the in-Service Education of Naval Officers, 1873–1973." Ph.D. diss., University of Delaware, 1974.

Herrick, Walter R. "General Tracy's Navy: A Study of the Development of American Sea Power, 1889–1893." Ph.D. diss., University of Virginia, 1962.

Herwig, Holger H. "The German Naval Officer Corps: A Social and Political History, 1890–1918." Ph.D. diss., State University of New York, 1973.

Jones, Jerry W. "U.S. Battleship Operations in World War I, 1917–1918." Ph.D. diss., University of North Texas, 1995.

Livermore, Seward W. "American Naval Development, 1898–1914: With Special Reference to Foreign Affairs." Ph.D. diss., Harvard University, 1943.

Miller, Craig D. "Rebuilding the U.S. Navy, 1865–1890." Ph.D. diss., Lamar University, 1979.

O'Connell, Robert L. "Dreadnaught? The Battleship, the United States Navy, and the World Naval Community." Ph.D. diss., University of Virginia, 1975.

Peterson, William S. "The Navy in the Doldrums: The Influence of Politics and Technology on the Decline and Rejuvenation of the American Fleet, 1866–1886." Ph.D. diss., University of Illinois, 1986.

Reardon, Carol Ann. "The Study of Military History and the Growth of Professionalism in the U.S. Army before World War I." Ph.D. diss., University of Kentucky, 1987.

Sexton, Donal J. "Forging the Sword: Congress and the American Naval Renaissance, 1880–1890." Ph.D. diss., University of Tennessee, 1976.

Stein, Stephen Kenneth. "Washington Irving Chambers: Innovation, Professionalization, and the New Navy, 1872–1919." Ph.D. diss., Ohio State University, 1999.

Wicks, Daniel Howard. "New Navy and New Empire: The Life and Times of John Grimes Walker." Ph.D. diss., University of California, Berkeley, 1979.

Wolters, Timothy Scott. "Managing a Sea of Information: Shipboard Command and Control in the United States Navy, 1899–1945." Ph.D. diss., Massachusetts Institute of Technology, 2003.

Websites

"Bigbadbattleships: Pre Dreadnaught Homeport."

U.S. Department of the Navy. "Dictionary of American Naval Fighting Ships." Naval Historical Center. http://www.history.navy.mil/danfs/.

INDEX

209

About the Author

Upon graduation from the U.S. Naval Academy in 1989, **Commander James C. "Chris" Rentfrow** completed flight school and was designated a Naval Flight Officer. After a career flying the EA-6B Prowler, Commander Rentfrow was selected to participate in the Permanent Military Professor program. He did his doctoral work at the University of Maryland, College Park and currently teaches U.S. and Naval History at the Naval Academy.